Animal Tracks
of Washington
and Oregon

ANIMAL TRACKS

OF

WASHINGTON

AND

OREGON

IAN SHELDON

© 1997 by Lone Pine Publishing
First printed in 1997   10   9   8   7
Printed in Canada

THE PUBLISHER: LONE PINE PUBLISHING

10145 – 81 Avenue	1808 – B Street NW, Suite 140
Edmonton, AB T6E 1W9	Auburn, WA 98001
Canada	USA

Website: www.lonepinepublishing.com

Canadian Cataloguing in Publication Data
Sheldon, Ian,
    Animal tracks of Washington and Oregon

    Includes bibliographical references and index.
    ISBN-13: 978-1-55105-090-0
    ISBN-10: 1-55105-090-0

    1. Animal tracks—Washington (State)—Identification. 2.
Animal tracks—Oregon—Identification. I. Title.
QL768.S532 1997   591.47'9   C97-910345-2

Senior Editor: Nancy Foulds
Editor: Roland Lines
Production Manager: David Dodge
Design, layout and production: Gregory Brown
Technical review: Donald L. Pattie
Animal Illustrations: Gary Ross, Horst Krause
Track Illustrations: Ian Sheldon
Cover Illustration: Gary Ross
Scanning: Elite Lithographers Ltd., Edmonton, Alberta, Canada

The publisher gratefully acknowledges the support of Alberta Community
Development and the Department of Canadian Heritage.

PC: P1

CONTENTS

INTRODUCTION

If you have ever spent time with an experienced tracker, or perhaps a veteran hunter, then you know just how much there is to learn about the subject and just how exciting the challenge of tracking animals can be. Maybe you think that tracking is no fun, because all you get to see are the animal's prints. What about the animal itself, isn't that much more exciting? Well, for most of us who don't spend a great deal of time in the beautiful wilderness of the Cascades, the chances of seeing the secretive Mountain Lion or the fun-loving River Otter are slim. The closest we may ever get to some animals will be through their tracks, and these can inspire a very intimate experience. Remember, you are following in the footsteps of the unseen—animals that are in pursuit of prey, or perhaps being pursued as prey.

This book offers an introduction to the complex world of tracking animals. Sometimes tracking is easy. Other times it is an incredible challenge that leaves you wondering just what animal left those unusual tracks. Take this book into the field with you, and it can provide some help with the first steps to identification. Prints and tracks are the book's focus; you will learn to recognize subtle differences for both. There are, of course, many additional signs to consider, such as scat and food caches, all of which help you to understand the animal you are tracking.

Remember, it takes many years to become an expert tracker. It is one of those skills that grows with you as you acquire new knowledge in new situations. Most importantly, you will have an intimate experience with nature. You will learn the secrets of the seldom seen. The more you discover, the more you will want to know, and by developing a good understanding of tracking you will gain an excellent appreciation of the intricacies and delights of our marvelous natural world.

How to Use This Book

First, take it into the field with you! Relying on your memory is not an adequate way to identify tracks. Tracking identification has to be done in the field, or with detailed sketches and notes that you can take home. Much of the process of identification is circumstantial, so you will have much more success when standing beside the track.

The book is laid out to be easy to use. There is a quick reference appendix on p. 140 to the tracks of all the animals illustrated in the book. This appendix is a fast way to familiarize yourself with certain tracks and the content of the book, and it guides you to the more informative descriptions of each animal and track.

Each animal description is illustrated with appropriate footprints and styles of track that they frequently leave. These illustrations are not exhaustive, but they show the most likely tracks or groups of prints that you will see. Where there are differences in orientation, left prints are illustrated. You will find a list of dimensions for the tracks, showing the general range, but there will always be extremes, just as there are with people who have unusually small or large feet. Under the category 'size' (of animal), the 'greater than' sign (>) is used when the size difference between sexes is pronounced.

If you think you may have identified a track, check the section on similar species for that animal. This section is designed to help you confirm your conclusions by pointing out other animals leave the same or similar tracks and showing you ways to distinguish between them.

As you read this book, you will notice an abundance of words like *often, mostly* and *usually.* Unfortunately, tracking will never be an exact science; we cannot expect animals to conform to our expectations, so be prepared for the unpredictable.

Tips on Tracking

As you flip through this guide, you will notice clear, well-formed prints. Do not be deceived! It is a rare track that will ever show so clearly. For a good, clear print the perfect conditions will be slightly wet, shallow snow that isn't melting, or slightly soft mud that isn't actually wet. Needless to say, these events can be rare, and most often you will be dealing with incomplete or faint prints, where you cannot really be sure of the number of toes.

Should you find yourself looking at a clear print, then the job of identification is much easier for you. There are a number of key features to look for. Measure the length and width of the print, count the number of toes, check for claw marks and note how far away they are from the body of the print, and look for a heel. Keep in mind other more subtle features like the spacing between toes and whether they are parallel or not, and whether there is fur on the sole of the print making it less clear.

Do not rely on the measurements of one print, but collect measurements from several prints to get an average impression. Even prints within one track can show a lot of variation.

When you are faced with the challenge of an unclear print—or even if you think you have made a successful identification from one print alone—it is time to think in the broader picture. Look beyond the single footprint and search out others.

Try to determine which is the fore print and which is the hind, and remember that many animals are built very differently from humans, having larger fore feet than hind feet. Sometimes the prints will overlap, or they can be directly on top of one another in a direct register. For some animals the fore and hind prints are pretty much the same.

Check out the pattern the prints make together in the track, and follow the trail for as many paces as is necessary for you to become familiar with the pattern. Patterns are very important, and can be the distinguishing feature between different animals.

Follow the trail for some distance, because it may give you some vital clues. For example, the trail may lead you to a tree, indicating that the animal is a climber, or it may lead down into a burrow. This can be the most rewarding part of tracking—you are following the life of the animal as it hunts, runs, walks, jumps, feeds or tries to escape a predator.

Take into consideration the habitat. Sometimes very similar species can only be distinguished by their location—one might be on the riverbank, while the other might be in the dense forest.

Think about your geographical location, too. Some animals have a limited range, perhaps only in remote parts of Washington and Oregon. This consideration can rule out some species and help you with your identification.

Remember that every animal will at some point leave a print or track that looks just like the track of a completely different animal!

Lastly, keep in mind that if you are quiet, you might catch up with the owner of the prints.

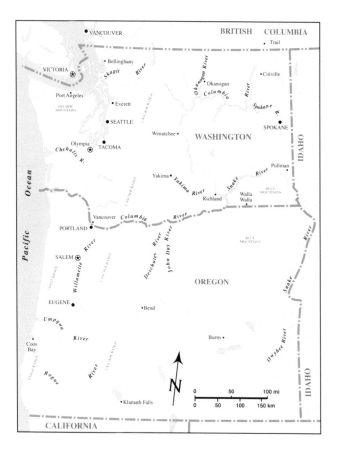

Terms and Measurements

Some of the terms used in tracking can be rather confusing, and they often depend on personal interpretation. For example, what comes to your mind if you see the word 'hopping'? Perhaps you see a person hopping about on one leg, or perhaps you see a rabbit hopping through the countryside. One person's perception of motion can be very different from another's. Below are some of these terms to clarify what is meant in this book, and where appropriate, how the measurement fits in with the term.

The following terms are sometimes used loosely and interchangeably. For example, while a rabbit might be described as a hopper, a squirrel seems to be more of a bounder, yet both leave a similar pattern of prints in the same sequence.

Bounding: Can be used interchangeably with hopping, jumping; often used for short distances; hind prints usually registering ahead of fore prints.

Galloping: Used for the motion made by animals with four even legs, such as dogs, at speed, hind prints registering ahead of fore.

Hopping: Usually tight clusters of prints, with two fore prints set between and behind the hind prints; similar to bounding.

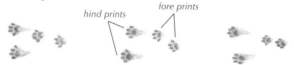

Running: Like galloping, but often applied generally to animals at speed.

Stotting: Confined to mule deer, describing the action of taking off from the ground at once, and landing on all four feet at once, in a pogo stick fashion.

Trotting: Faster than walking, slower than running.

Other tracking terms:

Alternating track: As made in walk by humans, left-right sequence, often a double register for four-legged animals, which are described as diagonal walkers.

double register   direct register

Dewclaws: Two small toe-like structures set above and behind the main foot of most hoofed animals.

> dew claws

\ dragline

Direct register: The hind print falls directly on the fore print.

Double register: The hind print falls slightly on or beside the fore print so that both can be seen at least in part.

Draglines: Lines left in snow or mud by foot or tail dragging over the surface.

Print: Fore and hind print treated individually; print dimensions are measured by length (including claws, maximum values may represent occasional heel register for some animals) and width; together prints make up a track.

Gallop group: Pattern made by prints in gallop, usually with hind prints registering ahead of fore prints, in a cluster of four.

Height: Taken at shoulder.

Length: Body length from head to rump, not including tail unless otherwise indicated.

Lope: Collection of four prints made at a fast pace, usually falling roughly in a line.

Retractable: Describing claws that can be pulled back in to keep them sharp, as with the cat family; claws do not register in print.

Sitzmark: Mark left on the ground by animal falling or jumping from a tree.

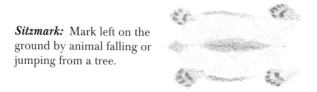

Straddle: Total width of the track, all prints considered.

Stride: Length from the center of one print to the center of the next print.

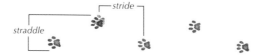

Track: Pattern left by a series of prints.

Trail: Often used to describe a track at length; think of it as the path of the animal.

MAMMALS

Fore and Hind Prints
Length: 3.3 in (8.4 cm)
Width: 2.5 in (6.4 cm)

Straddle
3.5–9 in (8.9–23 cm)

Stride
Walking: 8–19 in (20–48 cm)
Galloping:
    14 ft (4.3 m) or more

Size (buck>doe)
Height: 3 ft (91 cm)
Length: 3.8–4.9 ft
    (1.2–1.5 m)

Weight
75–135 lb (34–61 kg)

walking   *gallop group*

PRONGHORN ANTELOPE
Antilocapra americana

This gracious antelope frequents the wide-open grasslands and sagebrush plains. Pronghorns can be seen in south-eastern Oregon where they gather in groups of up to a dozen animals in summer, and as many as 100 animals in winter. Unlike deer, Pronghorns run for fun, easily attaining constant speeds of 40 mph (64 km/h), or as much as 60 mph (97 km/h) for short bursts.

A Pronghorn print has a pointed tip and a broad base. The hind prints usually register directly on top of the fore prints, making a tidy, alternating track. Pronghorns show a great tendency to drag their feet in snow, and they like to gather for feeding in areas where snow has been blown away. During their frequent gallops, the toe tips are spread wide, the dewclaws are absent, and the distance between gallop groups increases the faster the antelope moves.

Similar Species: Deer prints show dewclaws and have a narrower toe base, and deer tracks show a shorter stride between gallop groups; Mule Deer frequent similar habitats, but their tracks show a stot pattern.

19

Fore and Hind Prints
Length: 2.5–3.5 in (6.4–8.9 cm)
Width: 2–3.3 in (5.1–8.4 cm)
Straddle
6.5–12 in (17–30 cm)
Stride
Walking: 10–19 in (25–48 cm)
Size (billy>nanny)
Height: 3–3.5 ft (91–107 cm)
Length: 4.8–5.9 ft (1.5–1.8 m)
Weight
100–300 lb (45–135 kg)

walking

MOUNTAIN GOAT
Oreamnos americanus

Spotting the dazzling white coat of a Mountain Goat is truly a high-mountain, wilderness experience. Because of its preference for high terrain, this goat can only be seen in Olympic National Park, in the Cascade Mountains, and along the state border with Idaho into northeastern Oregon. The Mountain Goat's keen vision and its remote setting high above the treeline make it difficult to see; its tracks in snow may be your best clue that the goat is around.

The goat's tracks show it has long, widely spreading toes, which produce a squarish print. The hard rim and soft middle of the foot help this agile goat clamber over the most unlikely crags at remarkable speeds, but even the Mountain Goat can make a fatal mistake! In deeper snow the feet may leave draglines, and the dewclaws may register. The alternating walking track is a double register, hind over fore. If you are having difficulty identifying prints, the goat's preference for a remote habitat may be your best indicator that it is around; few deer or sheep climb so high. In severe weather, the Mountain Goat may come down into deer territory.

Similar Species: Deer or Bighorn Sheep prints are smaller, narrower and more pointed.

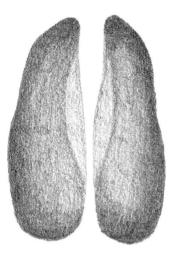

Fore and Hind Prints
Length: 2.5–3.5 in (6.4–8.9 cm)
Width: 1.8–2.5 in (4.6–6.4 cm)

Straddle
6–12 in (15–30 cm)

Stride
Walking: 14–24 in (36–61 cm)

Size (ram>ewe)
Height: 2.5–3.5 ft (76–107 cm)
Length: 4–6.5 ft (1.2–2 m)

Weight
75–275 lb (34–124 kg)

walking

BIGHORN SHEEP (Mountain Sheep)
Ovis canadensis

In late fall, the loud crack of two majestic rams head-butting one another can be heard for a great distance. To watch the rut is an awe-inspiring experience, but a rare one; this sheep prefers high, open meadows and scree slopes along the border mountains with Idaho, and only moves into valleys during winter. It tends to avoid forested areas, and is not quite as bold as Mountain Goats on craggy cliffs.

The print is quite squarish in shape and pointed towards the front. The outer edge of the hoof is hard while the inner part is soft, giving the sheep a good grip on tricky terrain. Find one track, and you will likely come across several, because this sheep likes to travel in herds. The neat, alternating walking pattern is a direct or double register of the hind over the fore. When a sheep runs, its toes will spread wide. The tracks may lead you to sheep beds—hollows dug into the snow that are used many times and often have a large accumulation of droppings.

Similar Species: Deer prints are more heart-shaped; Mountain Goat prints are wider at the toe, and Mountain Goats rarely run; domestic sheep prints are similar.

Fore and Hind Prints
Length: 2–3.3 in (5.1–8.4 cm)
Width: 1.6–2.5 in (4.1–6.4 cm)

Straddle
5–10 in (13–25 cm)

Stride
Walking: 10–24 in (25–61 cm)
Jumping: 9–19 ft (2.7–5.8 m)

Size (buck>doe)
Height: 3–3.5 ft (91–107 cm)
Length: 4–6.5 ft (1.2–2 m)

Weight
100–450 lb (45–203 kg)

walking   *stot group*

MULE DEER
Odocoileus hemionus

This widespread deer is frequently seen in meadows, open woodlands and desert plains. In winter, it moves down from higher terrain to warmer south-facing slopes and sagebrush flats, where it can still feed without having to contend with deep snow.

The Mule Deer has a neat alternating track, with the hind print registering on the fore print. It frequently uses the same well-worn trail in winter, and it prefers to stay in small groups. The prints are heart-shaped and sharply pointed. In deeper snow or when the animal is moving quickly, the prints will show dewclaws, which are closer to the hoofs on fore prints than on hind prints. Mule Deer have a unique gait for speed: they jump with all feet leaving and striking the ground at the same time—called 'stotting.' These stotting tracks will also show how the toes spread to distribute the weight and give better footing.

Similar Species: White-tailed Deer prints can be identical, but White-tails prefer denser cover and have a different gallop pattern with a shorter gallop stride; Pronghorn prints are wider at the base; Elk prints are longer and wider.

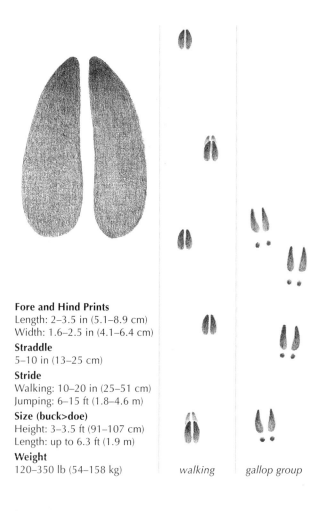

Fore and Hind Prints
Length: 2–3.5 in (5.1–8.9 cm)
Width: 1.6–2.5 in (4.1–6.4 cm)

Straddle
5–10 in (13–25 cm)

Stride
Walking: 10–20 in (25–51 cm)
Jumping: 6–15 ft (1.8–4.6 m)

Size (buck>doe)
Height: 3–3.5 ft (91–107 cm)
Length: up to 6.3 ft (1.9 m)

Weight
120–350 lb (54–158 kg)

walking   *gallop group*

WHITE-TAILED DEER
Odocoileus virginianus

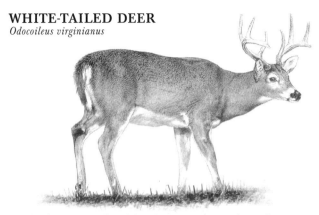

The keen eyesight of this deer guarantees that it knows about you before you know about it. Frequently, all we see is its flashing white tail as it gallops away, which earns this deer its other name of 'flagtail.' These adaptable deer may be found in small groups at the edges of forests and in brushlands, and they are widespread throughout Washington and in all but southeastern Oregon. White-tailed Deer can be common around ranches and residential areas.

The prints are heart-shaped and pointed, in an alternating track with direct- or double-registering hind prints on the fore prints. In snow, or when a deer is galloping on soft surfaces, the dewclaws register. This flighty deer gallops in the usual style, with the hind prints falling in front of fore prints and the toes spreading wide for better footing.

Similar Species: Mule Deer prints are almost identical, but the habitat or gallop prints help in diagnosis; Elk prints are longer and wider; Pronghorns prefer open spaces.

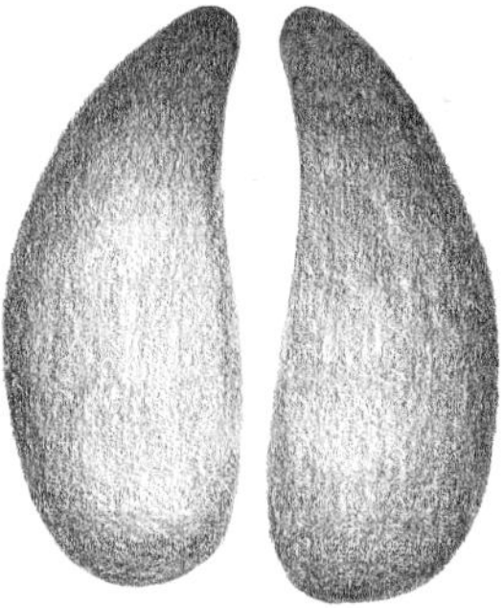

Fore and Hind Prints
Length: 3.2–5 in (8.1–13 cm)
Width: 2.5–4.5 in (6.4–11 cm)

Straddle
7–12 in (18–30 cm)

Stride
Walking: 16–34 in (41–86 cm)
Galloping: 3.3–7.8 ft (1–2.4 m)
Group length: to 6.3 ft (1.9 m)

Size (stag>hind)
Height: 4–5 ft (1.2–1.5 m)
Length: 6.5–10 ft (2–3 m)

Weight
500–1000 lb (225–450 kg)

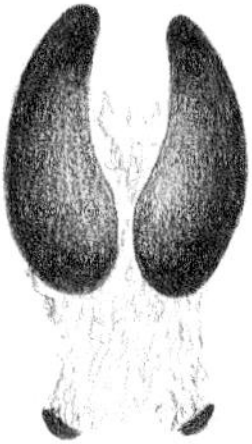

gallop print   *walking*

ELK (Wapiti)
Cervus elaphus

Elk are common in the Coast Ranges of both states, and they can also be found in parts of the Washington Cascades. Female elk are often seen in social herds, and they like to feed in forest openings and mountain meadows. Stags prefer to go solo, however, and they are easily recognized by their magnificent rack of antlers and distinctive bugling. Herds move into valleys when winter sets in.

Elk leave a neat alternating track with large, rounded prints, often in well-worn winter paths. The hind print will sometimes double register slightly ahead of the fore print. In deeper snow, or if an Elk gallops (with its toes spread wide), the dewclaws may register. A good place to look for Elk tracks is in the soft mud by summer ponds, where elk like to drink and sometimes splash around.

Similar Species: Deer prints can be similar but are generally smaller.

**Fore Print
(hind print is slightly smaller)**
Length: 4.5–6 in (11–15 cm)
Width: 4.5–5.5 in (11–14 cm)
Stride
Walking: 17–27 in (43–69 cm)

walking

HORSE
Equus caballus

This popular animal has unmistakable prints, but it deserves mentioning because you will come across its tracks in many places. Back-country use of the Horse means you can expect their tracks to show up almost anywhere.

Unlike any other animal discussed in this book, the Horse has only one huge toe, which leaves an oval print. A distinctive feature is the 'frog,' or V-shaped mark, at the base of the print. The horseshoe shows up clearly as a firm wall at the outside of the print. Not all horses will be shod, however, so don't expect to see this outer wall on every horse track. A typical, leisurely horse track is an alternating walk with the hind prints registering on or behind the slightly larger fore prints. Horses are capable of a range of speeds—up to a full gallop—but most recreational horseback riders take a more leisurely outlook on life.

Similar Species: Mule prints are smaller, and mules are rarely shod.

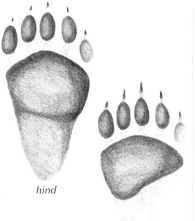

hind

fore

Fore Print
Length: 4–6.3 in (10–16 cm)
Width: 3.8–5.5 in (9.7–14 cm)

Hind Print
Length: 6–7 in (15–18 cm)
Width: 3.5–5.5 in (8.9–14 cm)

Straddle
9–14.5 in (23–37 cm)

Stride
Walking: 17–23 in (43–58 cm)

Size (male>female)
Height: 3–3.5 ft (91–107 cm)
Length: 5–6 ft (1.5–1.8 m)

Weight
200–600 lb (90–270 kg)

walking
(slow)

BLACK BEAR
Ursus americanus

The Black Bear is widespread in forested areas throughout both states. Finding fresh bear tracks can be a thrill, but take care; the bear may be just around the corner. Never underestimate the potential power of a surprised bear! Black Bears sleep deeply through winter, so don't expect to find their tracks in the colder months.

A Black Bear's print is about the size of a human print, but the bear's print is wider and shows claw marks. The small inner toe rarely registers. The fore foot has a small heel pad that often shows, and the hind foot has a big heel. The bear's slow walk results in a slightly pigeon-toed double register of the hind over the fore. More frequently, at a faster pace, the hind oversteps the fore, as shown for the Grizzly Bear. When the bear is running, the two hind feet register in front of the fore feet in an extended cluster. Along well-worn paths, look for digs—patches of dug-up earth—and bear trees—their scratched bark and claw marks show that these bears climb.

Similar Species: Grizzly prints are often larger, the toes are set more in a line and closer together, and the claws are longer.

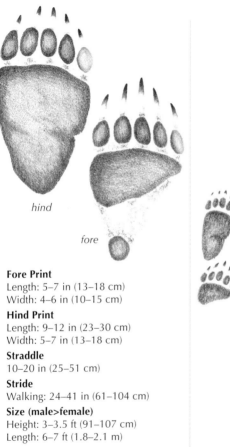

Fore Print
Length: 5–7 in (13–18 cm)
Width: 4–6 in (10–15 cm)

Hind Print
Length: 9–12 in (23–30 cm)
Width: 5–7 in (13–18 cm)

Straddle
10–20 in (25–51 cm)

Stride
Walking: 24–41 in (61–104 cm)

Size (male>female)
Height: 3–3.5 ft (91–107 cm)
Length: 6–7 ft (1.8–2.1 m)

Weight
200–850 lb (90–383 kg)

hind

fore

walking (fast)

GRIZZLY BEAR
Ursus arctos

The magnificent and imposing Grizzly Bear symbolizes mountain wilderness for many of us. It is very sparsely distributed and infrequently seen along the Washington border with Canada. Grizzly Bears prefer open country and valley bottoms, and they are sensitive to human activity. In winter, bears enter a deep slumber, so few of their tracks are seen.

One of their huge prints will show four or five toes, with very long claws and a small heel pad on the fore print. A solid rear heel makes for a sturdy hind print. The toes are closely set in a line, with the inner toe the smallest. The usual walking track shows the hind print registering ahead of the fore print, although a slower gait results in a track like the Black Bear's. The Grizzly occasionally gallops. Well-worn trails may lead to digs, trees with claw marks high up on the trunks or even the cache of a carcass. Take care if you find a cache, however, because the unpredictable Grizzly will likely be nearby.

Similar Species: Black Bear prints are smaller, with shorter claw marks and toes arranged more in an arc, and Black Bears make territorial tree scratchings lower on a tree trunk.

fore

hind

Fore Print
(hind print is slightly smaller)
Length: 4–5.5 in (10–14 cm)
Width: 2.5–5 in (6.4–13 cm)

Straddle
3–7 in (7.6–18 cm)

Stride
Walking: 15–32 in (38–81 cm)
Gallop: 3 ft (91 cm),
    leaps to 9 ft (2.7 m)

Size
(female is slightly smaller)
Height: 26–38 in (66–97 cm)
Length: 3.6–5.2 ft (1.1–1.6 m)

Weight
70–120 lb (32–54 kg)

walking    *trotting*

GRAY WOLF
(Timber Wolf)
Canis lupus

The soulful howl of a wolf epitomizes the outdoor experience, but it is seldom heard in either state. Your best bet for hearing a wolf is along the border with Canada. Reintroduction programs in a number of states are allowing the Gray Wolf to regain its former territory.

A wolf leaves a straight alternating track with the smaller hind print registering directly on the fore print. It is a large, oval print with all four claws showing. The heel pads of the fore prints have different lobing from those on the rear prints. If you find a track in the snow, it is likely the track of several wolves, because they will sensibly follow their leader through deep snow, sometimes dragging their feet. When a wolf trots, the hind print has a slight lead and falls to one side, giving an unbalanced appearance. Wolves and Coyotes gallop in the same way.

Similar Species: Domestic Dog (*Canis familiaris*) prints rarely can be as large, a dog's inner toes tend to spread out more, and the prints show a haphazard track where the register is not as direct; when a Wolverine's inner toe does not register, its print may be confused with a wolf's, but a Wolverine's pad shape is very different.

37

fore

hind

Fore Print
(hind print is slightly smaller)
Length: 2.4–3.1 in (6.1–7.9 cm)
Width: 1.6–2.4 in (4.1–6.1 cm)

Straddle
4–7 in (10–18 cm)

Stride
Walking: 8–16 in (20–41 cm)
Gallop: 2.5–10 ft (76 cm–3 m)

Size
(female is slightly smaller)
Height: 23–26 in (58–66 cm)
Length: 32–40 in (81–102 cm)

Weight
20–50 lb (9–23 kg)

walking   *gallop group*

COYOTE (Brush Wolf, Prairie Wolf)
Canis latrans

This widespread and adaptable canine prefers to hunt rodents and larger prey in open grasslands or woodlands. It hunts on its own, with a mate or as a family pack. A Coyote occasionally develops an interesting and cooperative relationship with a Badger, so you might find their tracks together after they have been digging for ground squirrels.

The oval fore prints are larger than the hind prints. Note the difference between the fore heel pad and the hind heel pad, which rarely registers clearly. Claw marks are usually only evident on the two center toes. Coyotes typically walk or trot in an alternating pattern, the walk having a wider straddle. The trail left by a Coyote is often very straight. A Coyote's tail hangs down, and in deep snow it leaves a dragline. When a Coyote gallops, the hind feet fall ahead of the fore feet; the faster the speed, the straighter the gallop group.

Similar Species: Domestic Dog (*Canis familiaris*) prints are not so oval and spread more, and a dog's trail is erratic and confused; Red Fox prints are usually, but not always, smaller; White-tailed Jackrabbit prints with no heel may look similar, but the gait is very different.

39

fore

hind

Fore Print
(hind print is slightly smaller)
Length: 1.3–2.1 in (3.3–5.3 cm)
Width: 1.1–1.5 in (2.8–3.8 cm)
Straddle
2–4 in (5.1–10 cm)
Stride
Walking/Trotting: 7–12 in (18–30 cm)
Size
Height: 14 in (36 cm)
Length: 21–29 in (53–74 cm)
Weight
7–15 lb (3.2–6.8 kg)

walking

GRAY FOX
Urocyon cinereoargenteus

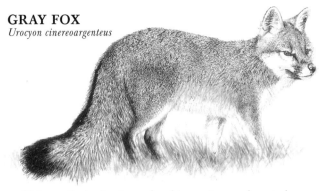

This small, shy fox is confined to western and central Oregon, preferring woodlands and chaparral. Follow its track, and you may be surprised to find this fox climbing a tree: it is the only fox to do so.

The fore print registers better than the hind print, and the long, semi-retractable claws may not always show on the hind print. The heel pads are often unclear—they sometimes show up just as small, round dots. This fox leaves a neat, alternating walking track; its trotting track is like a Red Fox's, and it has a gallop group like a Coyote's.

Similar Species: Swift Fox (*Vulpes velox*) prints are very similar but frequently smaller, and Swift Foxes prefer the arid regions of south Oregon; Red Fox prints show very different heel pads (with a bar across them) and are generally larger and less clear (because of thick fur), and Red Foxes generally have a longer stride and narrower straddle and are more widespread; Domestic Cat (*Felis catus*) or small Bobcat prints may be similar, but fox prints show claw marks and a smaller, more symmetrical heel pad.

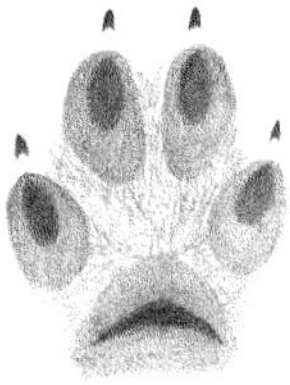

fore

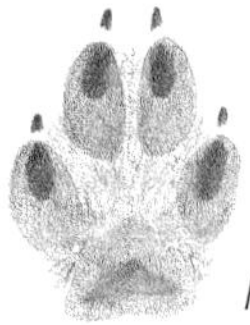

hind

Fore Print
(hind print is slightly smaller)
Length: 2.1–3 in (5.3–7.6 cm)
Width: 1.6–2.3 in (4.1–5.8 cm)

Straddle
2–3.5 in (5.1–8.9 cm)

Stride
Walking: 12–18 in (30–46 cm)
Trotting: 14–21 in (36–53 cm)

Size
(vixen is slightly smaller)
Height: 14 in (36 cm)
Length: 22–25 in (56–64 cm)

Weight
7–15 lb (3.2–6.8 kg)

walking    *trotting*

RED FOX
Vulpes vulpes

This beautiful and notoriously cunning fox is found in the western parts of both states, preferring mountainous forests and open areas. It is a very adaptable and intelligent animal.

The feet are very hairy, obscuring the finer details of the footprints and leaving only parts of the toes and heel pads showing. A very significant feature unique to this fox is the horizontal or slightly curved bar across the front heel pad. A Red Fox leaves a distinctive, straight trail of alternating prints made up of a direct register of the hind print on the wider fore print. When the fox trots, the hind print falls to one side of the fore print in a typically canid fashion. The fox gallops in the same manner as the Coyote, and the faster the gallop, the straighter the group of prints.

Similar Species: Domestic Dog (*Canis familiaris*) prints are of similar size but lack the bar on the heel pad, and dog tracks show a shorter stride with a less direct trail; small Coyote prints are similar, but with a wider straddle; Gray Fox prints can have some overlap in size, but Gray Foxes have a shorter stride and a wider straddle.

fore

hind

Fore Print
(hind print is slightly smaller)
Length: 1.8–2.5 in (4.6–6.4 cm)
Width: 1.8–2.6 in (4.6–6.6 cm)

Straddle
4–7 in (10–18 cm)

Stride
Walking: 8–16 in (20–41 cm)
Running: 4–8 ft (1.2–2.4 m)

Size
(female is slightly smaller)
Height: 20–22 in (51–56 cm)
Length: 25–30 in (64–76 cm)

Weight
15–35 lb (6.8–16 kg)

walking

*trotting
to loping*

BOBCAT (Wildcat)
Lynx rufus

The seldom-seen but adaptable Bobcat is widely distributed in both states, from wild mountainsides to chaparral, and even into residential areas. This stealthy hunter pursues its prey in the secret of the night.

The hind print is smaller and usually registers directly on the fore print in the walking track. The heel pads have two lobes in front and three to the rear. As a Bobcat picks up speed, its trail becomes a trot made of groups of two prints, with the hind print leading the fore. At even greater speeds, the trail becomes a group of four prints in a lope pattern. A Bobcat leaves draglines when it travels through deep snow. Unlike wild dogs, the trail meanders. If you follow this trail far enough, you may find half-buried scat—a sign of a Bobcat marking its territory.

Similar Species: Juvenile Mountain Lion or Lynx prints can be similar; large Domestic Cat (*Felis catus*) prints may be confused with juvenile Bobcat prints, but Domestic Cats have a shorter stride, narrower straddle and do not wander far from home, especially in winter; Fisher prints that don't show the fifth toe may look similar, so inspect the track for mustelid habits; dog, coyote and fox prints show claw marks, and the front of the footpad is lobed only once.

fore

hind

**Fore Print
(hind print is slightly smaller)**
Length: 3.5–4.5 in (8.9–11 cm)
Width: 3.5–4.8 in (8.9–12 cm)
Straddle
6–9 in (15–23 cm)
Stride
Walking: 12–28 in (30–71 cm)
Size
Length: 29–36 in (74–91 cm)
Weight
15–30 lb (6.8–14 kg)

walking

LYNX (Canada Lynx)
Lynx lynx

This large cat usually eludes humans by living in dense forests. The Lynx is sensitive to human interference, so it is abundant only in remote regions of Washington, and south into northeastern Oregon. With its huge feet and relatively light body weight, a Lynx stays on top of the snow while in pursuit of its main prey, the Snowshoe Hare.

The Lynx is a cautious walker that leaves an alternating track with neat direct registers of the hind print over the fore print. It rarely drags its feet in deeper snow. Thick fur on the feet obscures the prints' characteristics, and they often appear as big, round depressions with no detail. The print may be extended in deeper snow by 'handles' off to the rear. However deep the snow, this cat sinks no more than 8 inches (20 cm). A Lynx is more likely to bound than to run. Its curious nature results in a meandering trail that may lead you to a partially buried cache of food.

Similar Species: Bobcat prints are smaller; Mountain Lion prints are similar in size, but clearer, and Mountain Lions sink deeper in snow and have a wider straddle.

fore

hind

Fore Print
(hind print is slightly smaller)
Length: 3–4.3 in (7.6–11 cm)
Width: 3.3–4.8 in (8.4–12 cm)
Straddle
8–12 in (20–30 cm)
Stride
Walking: 13–32 in (33–81 cm)
Bounding: to 12 ft (3.7 m)
Size
Height: 26–31 in (66–79 cm)
Length: 3.5–5 ft (1.1–1.5 m)
Weight
70–200 lb (32–90 kg)

walking (fast)

MOUNTAIN LION
(Cougar)
Felis concolor

The Mountain Lion is shy, elusive and nocturnal in nature, so finding this cat's tracks is usually the best that trackers can hope for. Mountain lions are spread widely but sparsely because of their need for a big home territory.

The prints tend to be wider than long, and the retractable claws never register. In winter, thick fur on the feet makes a print look much larger, and it may obscure the two lobes on the front of the heel pad. In a walking track, the hind print will either direct or double register on the larger fore print. As the walking pace increases, the hind print will tend to fall ahead of the fore print. In snow, the thick, long tail may leave a dragline, which can obscure some of the footprints' details. A Mountain Lion seldom gallops, but when it needs to catch prey, it is capable of quick bounds.

Similar Species: Bobcat prints may be confused with juvenile Mountain Lion prints; Lynx prints are similar in size, but Lynx have a narrower straddle, a shorter stride and no tail drag, and they do not sink as deep in snow.

Fore and Hind Prints
Length: 1–1.4 in (2.5–3.6 cm)
Width: 1–1.4 in (2.5–3.6 cm)

Straddle
3–4 in (7.6–10 cm)

Stride
Walking: 3–6 in (7.6–15 cm)

Size
(female is slightly smaller)
Length: 24–32 in (61–81 cm)

Weight
1.5–2.5 lb (0.7–1.1 kg)

walking

RINGTAIL
(Cacomistle, Civet Cat, Miner's Cat)
Bassariscus astutus

This pretty cousin of the Raccoon is secretive and seldom seen, and it is only found in southwestern Oregon. The Ringtail is strictly nocturnal, and it rarely leaves any sign of its passage on the rocky terrain it frequents. It usually travels under the cover of shrubs, adding further to the difficulty of tracking this mammal. It is never too far from water.

The small, rounded prints show five toes, and the partially retractable claws will only occasionally register. A second heel pad is very rarely evident on the fore print, just behind the main pad. The common walking track is an alternating sequence of prints, with the hind print registering on, or close to, the fore print. A Ringtail's trail may lead you into rocky terrain, up a tree or to the animal's den.

Similar Species: Domestic Cat (*Felis catus*) prints only have four toes and never have a second heel pad; other small mustelid prints may be similar in size and shape, but Ringtails have a very different gait and habitat, and their fifth toe registers more often.

fore

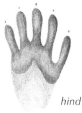

hind

Fore Print
Length: 2–3 in (5.1–7.6 cm)
Width: 1.8–2.5 in (4.6–6.4 cm)
Hind Print
Length: 2.4–3.8 in (6.1–9.7 cm)
Width: 2–2.5 in (5.1–6.4 cm)
Straddle
3.3–6 in (8.4–15 cm)
Stride
Walking/Running:
   7–20 in (18–51 cm)
Size (female is slightly smaller)
Length: 24–37 in (61–94 cm)
Weight
11–35 lb (5–16 kg)

walking   *running group*

RACCOON
Procyon lotor

The inquisitive Raccoon is adored for its distinctive face mask, yet disliked for its boundless curiosity—often demonstrated on garbage cans. Raccoons are common throughout both states. A good place to look for their tracks is near water at lower elevations.

The unusual Raccoon print looks like a human handprint, showing five well-formed toes. The small claws appear as dots. The highly dexterous forefeet rarely leave heel prints, but the hind prints, which are generally much clearer, do register a heel. The peculiar walking track shows the left fore print next to a right hind print and vice versa. The fore prints may be slightly ahead of the hind prints. Raccoons may also show an alternating track in deep snow, but this is rare because they den up in winter for the colder months. They occasionally run, with the hind prints falling ahead of the fore prints in a cluster. A Raccoon's trail may lead you up a tree, where Raccoons like to rest.

Similar Species: Fisher prints may be similar, but Fishers prefer deeper forests and show different gaits; unclear Opossum prints may look similar, but Opossums drag their tail.

fore

hind

Fore Print
Length: 2.5–3.5 in (6.4–8.9 cm)
Width: 2–3 in (5.1–7.6 cm)

Hind Print
Length: 3–4 in (7.6–10 cm)
Width: 2.3–3.3 in (5.8–8.4 cm)

Straddle
4–9 in (10–23 cm)

Stride
Walking/Running: 6–23 in (15–58 cm)

Size
(female is slightly smaller)
Length with tail: 3–4.3 ft (91–131 cm)

Weight
10–25 lb (4.5–11 kg)

running (fast)

RIVER OTTER
Lutra canadensis

No animal knows how to have more fun than a River Otter. If you are lucky enough to watch one at play, you will not soon forget the experience. Well-adapted for the aquatic environment, this otter is widespread along streams and other waterbodies. When you are in an otter's home territory, you are sure to find a wealth of evidence along the riverbanks. Otters are occasionally found in a forest, but they are usually going directly to another waterbody.

The five-toed prints show evidence of webbing in soft mud. The inner toes are set slightly apart. The fore prints do not show the webbing as clearly as the hind prints. The heel may register, lengthening the print. Otter tracks are very variable: they show the typical two-print bounding of mustelids, and, with faster runs, there are groups of four and three prints. The thick, heavy tail often leaves a mark over the prints. In snow, otters love to slide, often down riverbanks, leaving troughs nearly 12 inches (30 cm) wide. In summer they roll and slide on grass and mud.

Similar Species: Fisher prints are similar, but Fishers prefer forests and do not leave the conspicuous tail drag.

 55

fore

Fore Print
Length: 2.1–3.9 in
   (5.3–9.9 cm)
Width: 2.1–3.3 in
   (5.3–8.4 cm)

Hind Print
Length: 2.1–3 in (5.3–7.6 cm)
Width: 2–3 in (5.1–7.6 cm)

Straddle
3–7 in (7.6–18 cm)

Stride
Walking: 7–14 in (18–36 cm)
Run: 1–4.2 ft (30–128 cm)

Size (male>female)
Length with tail:
   34–41 in (86–104 cm)

Weight
3–12 lb (1.4–5.4 kg)

walking

running

FISHER (Black Cat)
Martes pennanti

This agile hunter is comfortable both on the ground and in the trees of mixed hardwood forests, more especially in the west of both states. Its speed and eager hunting antics make for exciting tracking—it races up trees and along the ground in its quest for squirrels. The Fisher's tracks may lead to the remains of an unfortunate Porcupine—the Fisher is one of the few predators to regularly kill and eat Porcupines.

Occasionally, a Fisher walks in a direct-registering alternating track, but it usually runs in the typical weasel fashion of an angled pair of prints, representing the direct register of hind over fore print. Five toes may register, but the small inner toe frequently does not. Only the fore foot has a small heel pad that can show up in the print. Tracks often vary within a short distance, and they are not associated with water: 'Fisher' is a misnomer!

Similar Species: Marten prints are usually much smaller (although male Marten prints may overlap in size with small female Fisher prints), and Martens weigh less, so they leave shallower prints.

Fore and Hind Prints
Length: 1.8–2.5 in
 (4.6–6.4 cm)
Width: 1.5–2.8 in
 (3.8–7.1 cm)

Straddle
2.5–4 in (6.4–10 cm)

Stride
Walking: 4–9 in (10–23 cm)
Running: 9–46 in (23–117 cm)

Size (male>female)
Length with tail:
 21–27 in (53–69 cm)

Weight
1.5–2.8 lb (0.7–1.3 kg)

walking   *running*

MARTEN (Pine Marten)

Martes americana

This aggressive predator is found in the coniferous montane and subalpine forests of both states. Size and habitat are often key to distinguishing a Marten's track from the very similar tracks of Fishers and Minks.

A Marten seldom leaves a clear print: the heel pad is very undeveloped, and in winter, the hairiness of the feet often obscures all pad detail, especially detail of the poorly developed palm pads. Prints often show only four toes, because the inner toe fails to register. The hind print registers on the fore print in an alternating walking pattern. In a bounding track, the hind prints fall on the fore prints in slightly angled groups of two, in the typical mustelid pattern. Gallop tracks may appear as clusters of three or four prints in a group. Follow the criss-crossing tracks, and you may find the Marten has scrambled up a tree; look for the sitzmark where it has jumped down again.

Similar Species: Female Fisher prints overlap in size with large male Marten prints, but Fishers leave clearer tracks; male Mink (*Mustela vison*) prints overlaps in size with small female Marten prints, but Mink do not climb trees and are often found near water.

Fore and Hind Prints
Length: 0.8–1.3 in (2–3.3 cm)
Width: 0.5–0.6 in (1.3–1.5 cm)
Straddle
1–2.1 in (2.5–5.3 cm)
Stride
Bounding: 9–35 in (23–89 cm)
Size (male>female)
Length with tail: 8–14 in
    (20–36 cm)
Weight
1–6 oz (28–170 g)

bounding

SHORT-TAILED WEASEL
(Ermine)

Mustela erminea

Weasels are active hunters with an avid appetite for rodents. Following their tracks can reveal much about these nimble creatures' activities. Although weasels are active all year, their tracks are most evident in winter, when they frequently burrow into the snow or, in pursuit of their prey, use an existing hole made by a rodent. Weasel tracks may lead you up a tree from time to time, and weasels have been known totake to water.

This weasel is smaller than a Long-tailed Weasel. It is widely distributed throughout Oregon and all but some central and eastern parts of Washington. It prefers woodlands and meadows up to higher elevations, not favoring wetlands or denser coniferous forests. Its bounding track may fall in clusters with alternating short and long strides.

Similar Species: Small female Long-tailed Weasel prints overlap in size with large male Short-tailed Weasel prints.

Fore and Hind Prints
Length: 1.1–1.8 in (2.8–4.6 cm)
Width: 0.8–1 in (2–2.5 cm)

Straddle
1.8–2.8 in (4.6–7.1 cm)

Stride
Bounding: 9.5–43 in
    (24–109 cm)

Size (male>female)
Length with tail:
    12–22 in (30–56 cm)

Weight
3–12 oz (85–340 g)

bounding

LONG-TAILED WEASEL
Mustela frenata

The typical trail of the weasel is a bounding pattern of paired prints. Because of a weasel's light weight and small, hairy feet, the pad detail is often obscured, especially in snow. Even with clear tracks, the inner toe rarely registers. Successful identification of the weasel track can be difficult. Your close attention to the straddle and stride is useful in tracking, but identification is further complicated by the fact that the small female of a larger species overlaps in size with the large male of a smaller species. Clues can be gained from the habit displayed in jumping patterns. Distribution and habitat may also be useful.

This weasel is larger than a Short-tailed Weasel, and it is more widely distributed throughout both states. Its tracks show typical bounding habits with an irregularity in the length of stride, which is sometimes short and sometimes long, with no consistent behavior. In deep snow, look for holes where the weasel has suddenly plunged into the snow, perhaps in pursuit of its prey.

Similar Species: Large male Short-tailed Weasel prints overlap in size with small female Long-tailed Weasel prints.

fore

hind

Fore Print
Length with heel: 4–7.5 in (10–19 cm)
Width: 4–5 in (10–13 cm)

Hind Print
Length: 3.5–4 in (8.9–10 cm)
Width: 4–5 in (10–13 cm)

Straddle
7–9 in (18–23 cm)

Stride
Walking: 3–12 in (7.6–30 cm)
Running: 10–40 in (25–102 cm)

Size
(female is slightly smaller)
Height: 16 in (41 cm)
Length: 32–46 in (81–117 cm)

Weight
18–47 lb (8.1–21 kg)

running (slow)

WOLVERINE
Gulo gulo

The robust and powerful Wolverine is the largest member of the weasel family, and its reputation has earned it many nicknames, such as 'skunk bear' and 'Indian devil.' Wolverines live in coniferous mountain forests, and their need for unadulterated wilderness has resulted in a scarce and scattered distribution in both states, although they may be spreading their range back into their former territory.

As with other mustelids, Wolverines have five toes, although the inner toe frequently does not register in the print. The fore print shows a small heel pad; the heel rarely registers on the hind print. Because of its low, squat shape, the Wolverine leaves a host of erratic trails, all typical of mustelids: an alternating walking pattern, the typical bounding pairs of prints or the common loping track of groups of three and four prints.

Similar Species: Wolverine tracks are more erratic and much larger than those of other mustelids.

fore

hind

**Fore Print
(hind print is slightly shorter)**
Length: 2.5–3 in (6.4–7.6 cm)
Width: 2.3–2.8 in (5.8–7.1 cm)

Straddle
4–7 in (10–18 cm)

Stride
Walking: 6–12 in (15–30 cm)

Size
Length: 21–35 in (53–89 cm)

Weight
13–25 lb (5.9–11 kg)

walking

BADGER
Taxidea taxus

   The squat shape and unmistakable face of this bold
animal are features suitable for the open grasslands, but
Badgers also venture into high mountain country. They are
found in the eastern regions of both states. Thick shoulders
and fore legs, coupled with long claws, make for a powerful
digging animal. Unlike other mustelids, the Badger likes to
den up in a hole for the really cold months of winter, so
look for the tracks in spring and fall snow.

   All five toes register on both pairs of feet. A Badger's
long claws are evident in the pigeon-toed track that it
leaves as it waddles along, although the claws on the hind
feet are not as long as on the fore feet. When a Badger
walks, the alternating track is a double register, with the
hind print sometimes falling just behind the fore print
and sometimes slightly ahead. The Badger's wide, low
body will often plow through deeper snow and obscure
the prints.

Similar Species: In snow, Porcupine tracks may be similar,
but they will show quill and tail draglines and will likely lead
up a tree, not to a hole.

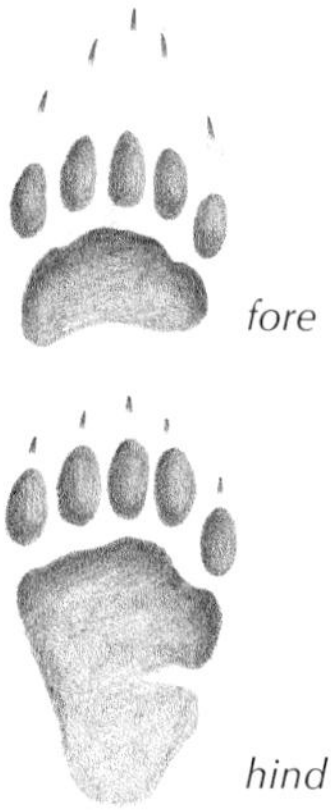

fore

hind

Fore Print
Length: 1.5–2.2 in
(3.8–5.6 cm)
Width: 1–1.5 in (2.5–3.8 cm)

Hind Print
Length: 1.5–2.5 in
(3.8–6.4 cm)
Width: 1–1.5 in (2.5–3.8 cm)

Straddle
2.8–4.5 in (7.1–11 cm)

Stride
Walking/Running: 2.5–8 in
(6.4–20 cm)

Size
Length with tail: 20–32 in
(51–81 cm)

Weight
6–14 lb (2.7–6.3 kg)

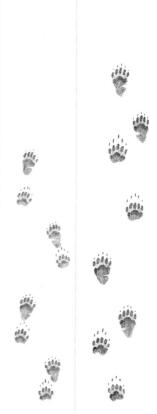

walking (fast)   *running*

STRIPED SKUNK
Mephitis mephitis

This striking skunk has a notorious reputation for its vile smell; the lingering odor is often the best sign of its presence. Striped Skunks den up in winter, coming out on warmer days and in spring. They are widespread throughout both states, preferring lower elevations and diverse habitats.

Both feet have five toes, and the long claws on the fore feet often register. The smooth palm pads and small heel pads leave surprisingly small prints. Skunks mostly walk; with such a potent smell for their defense, and those memorable black and white stripes, they rarely need to run. They leave a track that rarely shows any consistent pattern—a distinctive feature of this mustelid—but an alternating walking pattern may be evident. The greater the speed, the more the hind print oversteps the fore. Should a skunk need to run, its track is a pattern of clumsy, four-print groups set closely to each other. Striped Skunks drag their feet in snow.

Similar Species: Western Spotted Skunks have smaller prints, in a very random pattern, and a more limited range along the coast of Washington and most of Oregon.

fore

hind

Fore Print
Length: 1–1.3 in (2.5–3.3 cm)
Width: 0.9–1.1 in (2.3–2.8 cm)

Hind Print
Length: 1.2–1.5 in (3–3.8 cm)
Width: 0.9–1.1 in (2.3–2.8 cm)

Straddle
2–3 in (5.1–7.6 cm)

Stride
Walking: 1.5–3 in (3.8–7.6 cm)
Jumping: 6–12 in (15–30 cm)

Size
Length: 13–25 in (33–64 cm)

Weight
0.6–2.2 lb (0.3–1 kg)

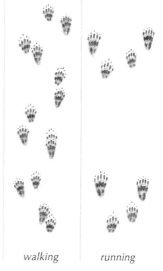

walking   *running*

WESTERN SPOTTED SKUNK (Civet Cat)

Spilogale gracilis

This beautifully marked skunk is smaller than its striped cousin and occurs in the western parts of both states, and especially in Oregon, in diverse habitats, such as scrubland, forests and farmland. Its nocturnal habits make it a rare sight, however, and its tracks are rare in winter—skunks den up to avoid the coldest months. It is less apt to spray than a Striped Skunk, so its odor is less frequently detected.

This skunk leaves a very haphazard trail as it forages for food on the ground and occasionally climbs trees with ease. The long claws on the fore feet often register, and the palm and heel may show some defined pads. On the rare occasions that a Spotted Skunk runs, it may bound along, leaving groups of four prints, hind ahead of fore.

Similar Species: Striped Skunks are bigger, with larger prints, they have less scattered tracks, a shorter running stride (or they jump) and a more widespread distribution in Washington, and they do not climb trees.

71

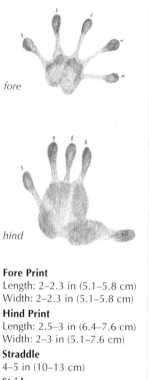

fore

hind

Fore Print
Length: 2–2.3 in (5.1–5.8 cm)
Width: 2–2.3 in (5.1–5.8 cm)

Hind Print
Length: 2.5–3 in (6.4–7.6 cm)
Width: 2–3 in (5.1–7.6 cm)

Straddle
4–5 in (10–13 cm)

Stride
5–11 in (13–28 cm)

Size
Length: 2–2.5 ft (61–76 cm)

Weight
9–13 lb (4.1–5.9 kg)

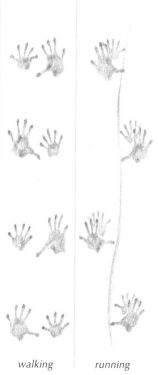

walking   *running*

OPOSSUM
Didelphis virginiana

This slow-moving, nocturnal marsupial is found along the western coastal regions of both states. While it is found in many types of habitat, it prefers open woodland or brushland around water. It is quite tolerant of residential areas. Opossum tracks can often be seen in mud by water and in snow during the warmer months of winter—Opossums tend to den up during severe weather.

Opossums are excellent climbers, so do not be surprised if their tracks lead to a tree. They have two walking habits: the common alternating pattern, with the hind print registering on the fore print, or the Raccoon-like pairs of prints, with the hind print next to opposing fore print. The long, inward pointing thumb of the hind foot is very distinctive, and it does not show a claw mark. In snow, the dragline of the long, naked tail may be stained with blood; unfortunately, this thinly haired animal frequently suffers frostbite.

Similar Species: The thumb is very distinctive, but if the print is unclear it may be confused with a Raccoon print.

fore

hind

Fore Print
Length: 2.5–4 in (6.4–10 cm)
Width: 1.3–1.7 in (3.3–4.3 cm)
Hind Print
Length: 3.5–6.5 in (8.9–17 cm)
Width: 1.5–2.5 in (3.8–6.4 cm)
Straddle
4.5–7 in (11–18 cm)
Stride
Hopping: 1–10 ft (30 cm–3 m)
Size
Length: 23–25 in (58–64 cm)
Weight
5–9 lb (2.3–4.1 kg)

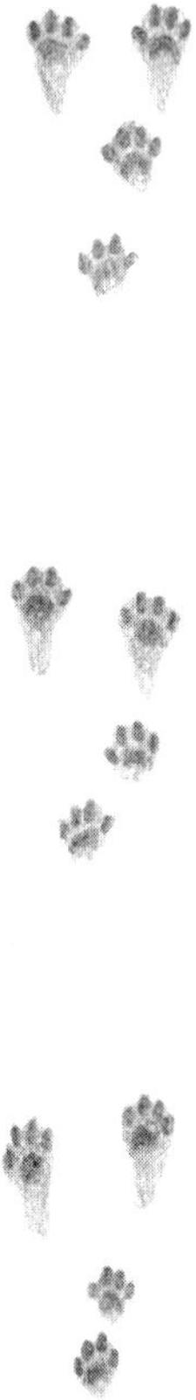

hopping

WHITE-TAILED JACKRABBIT
Lepus townsendii

This hare frequents the open country of the eastern parts of both states. An athletic animal, it can reach speeds of 45 mph (72 km/h). It is infrequently seen because of its nocturnal and solitary habits.

Both fore and hind prints show four toes, and the hind foot may often register a long heel. This jackrabbit's hopping action creates triangular print groups that spread out considerably as the hare speeds up. With its strong hind legs, it is capable of leaping as much as 20 feet (6.1 m) to escape predators. Following the tracks could lead you to its 'form'—a depression where it rests—or they may reveal an urgent zigzag pattern, indicating that the hare had to flee from danger.

Similar Species: Large Black-tailed Jackrabbit (*Lepus californicus*) prints may be the same size, but White-tailed Jackrabbits can be found further north than Black-taileds, which are common along the Oregon coast; Snowshoe Hare hind prints are much bigger, and Snowshoe Hares are not capable of such big leaps and require dense cover.

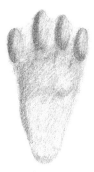

fore

hind

Fore Print
Length: 2–3 in (5.1–7.6 cm)
Width: 1.5–2 in (3.8–5.1 cm)
Hind Print
Length: 4–6 in (10–15 cm)
Width: 2–3.5 in (5.1–8.9 cm)
Straddle
6–8 in (15–20 cm)
Stride
Hopping: 10 in–4.2 ft (25–128 cm)
Size
Length: 12–21 in (30–53 cm)
Weight
2–4 lb (0.9–1.8 kg)

hopping

SNOWSHOE HARE
(Varying Hare)
Lepus americanus

This hare is well known for its color change, from summer brown to winter white, and for its huge hind feet, which enable it to 'float' on the surface of snow. Widespread in all but some parts of eastern Oregon, it frequents brushy areas in forests, which provide good cover from Lynx and Coyotes, its most likely predators. Hares are most active at night.

Like all rabbits, the Snowshoe Hare's most common track is a hopping one, with groups of four prints in a triangular pattern. These groups can be quite long if the hare is running quickly. The hind prints are much larger than the fore prints. In winter, heavy fur on the hind feet thickens the toes, which are able to splay out, further spreading the hare's weight when it runs on snow. You might find well-worn runways that are often used as escape runs. If you are lucky, you might even come across a resting hare, since they do not live in burrows. Other signs of their presence are twigs and stems that have been severed neatly at a 45° angle.

Similar Species: Cottontail prints are smaller; open country jackrabbits are larger, but they have smaller hind prints.

fore

hind

Fore Print
Length: 1–1.5 in (2.5–3.8 cm)
Width: 0.8–1.3 in (2–3.3 cm)
Hind Print
Length: 3–3.5 in (7.6–8.9 cm)
Width: 1–1.5 in (2.5–3.8 cm)
Straddle
4–5 in (10–13 cm)
Stride
Hopping: 7–36 in (18–91 cm)
Size
Length: 12–17 in (30–43 cm)
Weight
1.3–3 lb (0.6–1.4 kg)

hopping

MOUNTAIN COTTONTAIL
(Nuttall's Cottontail)
Sylvilagus nuttallii

This abundant rabbit is found widely through the eastern regions of both states. It prefers sagebrush, but it can be found in grasslands and rockier areas, where it will hide in dense vegetation and crevices to avoid predators, such as Bobcats and Coyotes.

As with other rabbits, its most common track is the triangular grouping of four prints, with the larger hind prints falling ahead of the fore prints. The two fore prints might overlap. The hairy toes will obscure any pad detail you might otherwise hope for. The hind print can often appear rather pointed. Following the tracks, you could be startled if the rabbit flies out from its 'form' (a depression in the snow or ground where it rests).

Similar Species: Jackrabbits have much larger prints and a longer stride; squirrel tracks show a similar pattern, but the fore prints are more consistently side by side.

fore

hind

Fore Print
Length: 0.8 in (2 cm)
Width: 0.6 in (1.5 cm)
Hind Print
Length: 1–1.2 in (2.5–3 cm)
Width: 0.6–0.8 in (1.5–2 cm)
Straddle
2.5–3.5 in (6.4–8.9 cm)
Stride
Walking/Running: 4–10 in (10–25 cm)
Size
Length: 6.5–8.5 in (17–22 cm)
Weight
4–6 oz (113–170 g)

hopping

PIKA (Cony, Rock Rabbit)
Ochotona princeps

When hiking high up in the mountains, you are more likely to hear the squeak of this cousin of the rabbit than see one. The Pika, confined to areas of high elevation, rarely leaves good tracks because it usually travels across exposed, rocky areas on mountain slopes. Although they are active during winter, Pikas stay under the snow and feed on stored food. In summer, they are quick to disappear under rocks for protection.

Pika tracks may be found in spring on patches of snow or mud. The fore print shows five toes, although the fifth toe may not register, while the hind print has only four toes. The prints may appear in an erratic alternating pattern or in three- and four-print bounding groups. The Pika's little hay piles, which dry in the sun in preparation for the long winter ahead, are more conspicuous signs of its presence.

Similar Species: Pika prints are rarely confused with other rabbit prints because of the high-mountain habitat; Mountain Cottontail fore prints never show five toes, and the hind prints have a long heel.

fore

hind

Fore Print
Length: 2.3–3.3 in (5.8–8.4 cm)
Width: 1.3–1.9 in (3.3–4.8 cm)
Hind Print
Length: 2.8–3.9 in (7.1–9.9 cm)
Width: 1.5–2 in (3.8–5.1 cm)
Straddle
5.5–9 in (14–23 cm)
Stride
Walking: 5–10 in (13–25 cm)
Size
Length with tail: 26–41 in (66–104 cm)
Weight
10–28 lb (4.5–13 kg)

walking

PORCUPINE
Erethizon dorsatum

This notorious rodent rarely runs, because its many long quills are a formidable defense. Widespread throughout both states, the Porcupine shows a preference for forests, but it can also be seen in more open areas.

The most common Porcupine track is an alternating walking pattern, with the longer hind print registering on or slightly ahead of the fore print. Look for long claw marks on both prints. Note that the fore print only has four toes, not five, as is often described. On clear prints, the unusual pebbly surface of the solid heel pads may show. It is evident from its track that the Porcupine has a waddling gait, and its pigeon-toed footprints are often obscured by scratches from its heavy, spiny tail. In deeper snow, this squat animal drags its feet, and it may leave a trough with its body. A Porcupine's trail might lead you to a tree, where these animals spend much of their time feeding—look for chewed bark or nipped buds lying on the forest floor.

Similar Species: Badgers have pigeon-toed prints, but their tracks don't show tail drag, and Badgers don't climb trees.

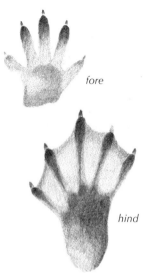

fore

hind

Fore Print
Length: 2.5–4 in (6.4–10 cm)
Width: 2–3.5 in (5.1–8.9 cm)

Hind Print
Length: 5–7 in (13–18 cm)
Width: 3.3–5.3 in (8.4–13 cm)

Straddle
6–11 in (15–28 cm)

Stride
Walking: 3–6.5 in (7.6–17 cm)

Size
Length with tail: 3–3.9 ft (91–119 cm)

Weight
28–75 lb (13–34 kg)

walking

BEAVER
Castor canadensis

Few animals leave as many signs of their presence as the Beaver, the largest North American rodent and a common sight around water. Common signs of Beaver activity are the very evident dams and lodges and stumps of felled trees; trunks gnawed clean of bark show marks from the Beaver's huge incisors. Another sign of recent Beaver activity is a scent mound marked with castoreum, which is a yellowish, strong-smelling fluid that Beavers produce.

A Beaver's tracks are often obscured by its thick, scaly tail or by the branches it drags about for construction and food, but if you find tracks where individual prints are evident, look for webbing on the large hind prints. Broad toenails, with the inner second toenail usually not showing, are another feature of the hind print. Although Beavers have five toes on each foot, it is rare for all of them to register. The track may be in an irregular alternating sequence, often with a double register. Beavers often use the same path, resulting in a well-worn trail.

Similar Species: Little confusion arises, because Beavers leave many signs of their presence.

fore

hind

Fore Print
Length: 1.1–1.5 in (2.8–3.8 cm)
Width: 1.1–1.5 in (2.8–3.8 cm)

Hind Print
Length: 1.6–3.2 in (4.1–8.1 cm)
Width: 1.5–2.1 in (3.8–5.3 cm)

Straddle
3–5 in (7.6–13 cm)

Stride
Walking: 3–5 in (7.6–13 cm)
Running: to 1 ft (30 cm)

Size
Length with tail: 16–25 in (41–64 cm)

Weight
2–4 lb (0.9–1.8 kg)

walking

MUSKRAT
Ondatra zibethica

Like the Beaver, this rodent is found through-out both states wherever there is water. Beavers are very tolerant of Muskrats and even allow them to live in parts of their lodges. Muskrats are active all year, and they leave plenty of signs of their presence. They have an extensive network of burrows, often undermining the riverbank, so do not be surprised if you suddenly fall into a hidden hole! Other signs of this rodent are the small lodges in water, and beds of vegetation on which they rest, sun and feed during summer.

The fore feet each have five toes, but the inner toe is reduced and rarely registers in the print. The hind print shows five well-formed toes. The prints are usually in an alternating track, with the hind print just behind or slightly overlapping the fore print. In snow, a Muskrat drags its feet a lot, and its tail leaves a sweeping dragline.

Similar Species: Few animals share this water-loving rodent's habits.

fore

hind

Fore and Hind Prints
Length: 1.6–2 in (4.1–5.1 cm)
Width: 1 in (2.5 cm)

Straddle
2–3 in (5.1–7.6 cm)

Stride
Walking: 3 in (7.6 cm)

Size
Length: 9.5–18.5 in (24–47 cm)

Weight
1–3 lb (0.5–1.4 kg)

walking

MOUNTAIN BEAVER
(**Aplodontia**)

Aplodontia rufa

This stocky rodent is a curious resident of the dark, moist forests of western Oregon and Washington. Its tail is so short that it is rarely seen. The Mountain Beaver is neither a beaver nor a montane animal; it is an enthusiastic burrower that leaves a labyrinth of shallow tunnels. Its burrowing activity can easily be seen after the snow melts to reveal soil cores much like those of a Pocket Gopher, only larger.

Mountain Beavers leave narrow prints, with the hind print showing a slightly longer heel, in an alternating walking track. The hind print occasionally registers on the fore print. Both feet have five toes, but the inner toe does not fully register on the fore print. Mountain Beavers are slow-moving animals that like to stay near home, so you are more likely to notice other signs of their activity, such as the large haymaking piles left on logs or on the ground. Look around for the burrows nearby. On the trees and shrubs, you may notice the ends of the shoots neatly nipped off; sometimes its appetite for clipping leaders or peeling bark off conifers can cause serious damage to trees.

Similar Species: Few animals of dark, moist forests share the unusual prints and activities of a rodent this size.

 89

fore

hind

Fore and Hind Prints
Length: 1.5–2.5 in (3.8–6.4 cm)
Width: 1–1.5 in (2.5–3.8 cm)
Straddle
3–5 in (7.6–13 cm)
Stride
Walking: 2–6 in (5.1–15 cm)
Running: 3–14 in (15.4–36 cm)
Size (male>female)
Length with tail:
17–28 in (43–71 cm)
Weight
5–10 lb (2.3–4.5 kg)

walking   *running*

YELLOW-BELLIED MARMOT
Marmota flaviventris

These endearing squirrels seem to have a good life: sleeping all winter and sunbathing on rocks in summer. Yellow-bellied Marmots are found from the foothills to high rocky areas in the central and eastern regions of both states. They live in small colonies, with an extensive network of burrows, and they are a joy to watch when they play-fight.

The fore print shows four toes and three palm pads, and sometimes two heel pads. The hind print shows five toes, four palm pads and two poorly registering heel pads. The hind print registers over the fore print in the usual alternating walking pattern, but when a marmot runs, it will leave a group of four prints, with the hind prints ahead of the fore. Because of marmots' preference for rocky habitats, their tracks can be hard to find.

Similar Species: Hoary Marmot (*Marmota caligata*) prints are longer than 2.5 in (6.4 cm), and this larger marmot, which likes similar high-altitude habitats, is found in the central and northern regions of Washington, overlapping the Yellow-bellied Marmot's range; small Raccoons may leave similar running tracks, but a Raccoon shows five toes on each fore print.

fore

hind

Fore Print
Length: 1–1.3 in (2.5–3.3 cm)
Width: 0.5–1 in (1.3–2.5 cm)

Hind Print
Length: 1.1–1.5 in (2.8–3.8 cm)
Width: 0.8–1.3 in (2–3.3 cm)

Straddle
2.3–4 in (5.8–10 cm)

Stride
Running: 7–20 in (18–51 cm)

Size
Length with tail: 8–13 in (20–33 cm)

Weight
6–10 oz (170–284 g)

running

GOLDEN-MANTLED GROUND SQUIRREL
Spermophilus lateralis

The Golden-mantled Ground Squirrel is one of the more widespread of the several species of ground squirrel in Washington and Oregon, and it enjoys varying terrain, from open forests to rocky areas above treeline. These bold squirrels can be very tame when they beg for food, but don't give in; our unhealthy diet certainly doesn't suit them!

Ground squirrels hibernate during winter; their tracks are most evident after late or early snowfalls or in mud around their many burrow entrances. The fore foot has such a reduced fifth toe that it rarely registers, and the two heel pads sometimes show. The larger hind foot has five toes. Both fore and hind feet have long claws that often show in the prints. Ground squirrels are usually seen scurrying around, leaving a typical squirrel track—hind prints registering ahead of diagonally placed fore prints.

Similar Species: California Ground Squirrels (*Spermophilus beecheyi*) are larger and are found in western Oregon; Columbian Ground Squirrels (*S. columbianus*) are found along the eastern borders of both states; chipmunk tracks are smaller; tree squirrel tracks have a more square-shaped running group.

fore

hind

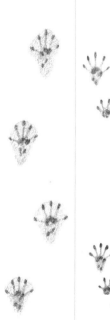

Fore Print
Length: 0.6–0.8 in (1.5–2 cm)
Width: 0.4–0.5 in (1–1.3 cm)

Hind Print
Length: 1–1.5 in (2.5–3.8 cm)
Width: 0.6–0.8 in (1.5–2 cm)

Straddle
2.3–2.7 in (5.8–6.9 cm)

Stride
Walking: 1.8–3 in (4.6–7.6 cm)
Jumping: 5–8 in (13–20 cm)

Size
Length with tail:
    11–19 in (28–48 cm)

Weight
7–21 oz (198–595 g)

walking   *running*

BUSHY-TAILED WOODRAT (Packrat)

Neotoma cinerea

The Bushy-tailed Woodrat is found almost everywhere in both states, especially in rugged terrain and forests, but it is not as fond of deserts as other woodrat species. Tracking one of these nocturnal rodents can be very rewarding; the trail might lead you to its distinctive, massed nest, which can be 5 feet (1.5 m) across and is sometimes in an abandoned building. This animal is a curious hoarder that brings home all manner of objects—it serves as a rather selective wilderness garbage collector.

Woodrats often walk in an alternating fashion, with direct register of the hind print on the fore. The stride is short relative to the size of the prints. Four toes show on the fore print and five on the hind. The short claws rarely show. This woodrat frequently runs, leaving a pattern of four prints, with the larger hind prints registering ahead of the diagonally placed fore prints. Do not be surprised if its trail ends at the base of a tree.

Similar Species: Norway Rats are usually found close to human activity; marmot prints are similar but much larger; other less abundant species of woodrat are similar.

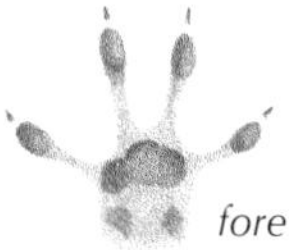

fore

hind

Fore Print
Length: 0.7–0.8 in (1.8–2 cm)
Width: 0.5 in (1.3 cm)

Hind Print
Length: 1–1.3 in (2.5–3.3 cm)
Width: 0.8–1 in (2–2.5 cm)

Straddle
3 in (7.6 cm)

Stride
Walking: 1.5–3.5 in (3.8–8.9 cm)
Jumping: 5–12 in (13–30 cm)

Size
Length with tail:
    12.5–18.5 in (32–47 cm)

Weight
7–18 oz (198–510 g)

walking

NORWAY RAT
Rattus norvegicus

This despised rat is widespread almost anywhere humans have decided to build homes, although it is not entirely dependent on people and may live in the wild as well.

Active both day and night, this colonial rat leaves groups of four prints when it runs, with the hind prints in front of the diagonally placed fore prints. The fore print shows four toes, while the hind print shows five toes. More commonly, rats leave an alternating walking pattern, with the larger hind prints close to or overlapping the fore prints, and the hind heel does not show. In snow the tail often leaves a dragline. Rats live in groups, so you may find many tracks together, often leading to their 2-inch (5.1 cm) wide burrows.

Similar Species: Woodrat tracks may be similar, but woodrats rarely associate with human activity, except in abandoned buildings.

fore

hind

walking

Fore Print
Length: 1 in (2.5 cm)
Width: 0.6 in (1.5 cm)
Hind Print
Length: 0.8–1 in (2–2.5 cm)
Width: 0.5 in (1.3 cm)
Straddle
1.5–2 in (3.8–5.1 cm)
Stride
Walking: 1.3–2 in (3.3–5.1 cm)
Size (male>female)
Length: with tail: 6–9 in (15–23 cm)
Weight
2.8–5 oz (79–142 g)

NORTHERN POCKET GOPHER
Thomomys talpoides

This seldom-seen rodent is found in eastern regions of both states, from open pine forests to high alpine meadows. It spends most of its time in burrows, only venturing out to move mud around and to find a mate. Because of their need for digging, pocket gophers prefer soft, moist soils.

By far, the best signs of Northern Pocket Gopher activity are the muddy mounds and tunnel cores that are especially evident after the spring thaw. A mound will mark the entrance to the burrow, and it is always blocked up with a plug. Search around the mound and you may find prints. Both feet have five toes, and the front foot has well-developed, long claws for digging, although it is a rare print that shows this much detail. Pocket gophers typically walk in an alternating track, with the hind print registering on or slightly behind the fore print.

Similar Species: Northern Pocket Gopher tracks are associated with its distinctive burrows, leaving little room for confusion, except with other less abundant pocket gopher species.

fore

hind

Fore Print
Length: 0.5–0.8 in (1.3–2 cm)
Width: 0.5 in (1.3 cm)

Hind Print
Length: 1.3–1.8 in (3.3–4.6 cm)
Width: 0.8 in (2 cm)

Straddle
3–3.8 in (7.6–9.7 cm)

Stride
Running: 11–29 in (28–74 cm)

Size
Length with tail: 9–11.5 in (23–29 cm)

Weight
4–6.5 oz (113–184 g)

sitzmark into running

NORTHERN FLYING SQUIRREL
Glaucomys sabrinus

These acro-
bats are found
in coniferous
forests at quite
high altitudes, in
all but the central and
southern regions of both states. They prefer widely spaced
forests, where they can make the most of the membranous
flap between their legs and glide through the night, from
tree to tree. Northern Flying Squirrels will den up together
in a tree cavity for warmth.

Because of its gliding, this squirrel does not leave as
many tracks as other squirrels, and it is very difficult to
find any evidence of its presence in summer. In winter,
however, you might come across a sitzmark–the distinctive
pattern a squirrel leaves when it lands in the snow–and a
short bounding track, as it rushes off to the nearest tree or
to do some quick foraging. This squirrel's tracks are typical
of squirrels and other rodents, with the hind prints falling
ahead of the fore prints, and the fore prints usually registering side by side.

Similar Species: Red Squirrel prints are usually larger,
and Red Squirrels never leave a sitzmark, but in deep snow
when the tracks are unclear, it can be impossible to tell
these two squirrels apart; chipmunks have a smaller
straddle and smaller prints.

 101

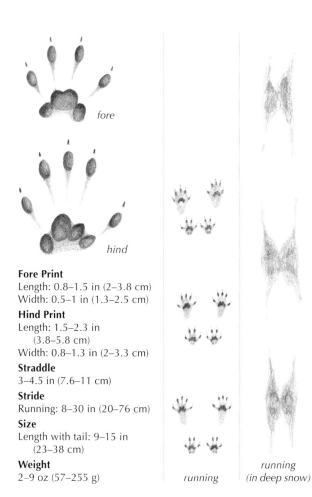

fore

hind

Fore Print
Length: 0.8–1.5 in (2–3.8 cm)
Width: 0.5–1 in (1.3–2.5 cm)

Hind Print
Length: 1.5–2.3 in
  (3.8–5.8 cm)
Width: 0.8–1.3 in (2–3.3 cm)

Straddle
3–4.5 in (7.6–11 cm)

Stride
Running: 8–30 in (20–76 cm)

Size
Length with tail: 9–15 in
  (23–38 cm)

Weight
2–9 oz (57–255 g)

running

*running
(in deep snow)*

RED SQUIRREL
(Pine Squirrel, Chickaree)
Tamiasciurus hudsonicus

When you enter Red Squirrel territory, this squirrel greets you with a loud, chattering call. It can be found in northern and eastern regions of both states. Other obvious signs of this widespread forest dweller are large middens—piles of cone scales and cores left at the bottom of trees—which indicate a squirrel's favorite feeding site.

Active year-round in their small territories, Red Squirrels leave an abundance of tracks that lead from tree to tree or down a burrow. These energetic animals run mostly, in a gait that leaves groups of four prints, hind falling ahead of fore. The fore prints tend to be side by side, although this is not always the case. Four toes show on each fore print, and five on each hind print. The heels do not often register when squirrels run. In deeper snow, the prints merge to form pairs of diamond-shaped tracks.

Similar Species: The larger Western Gray Squirrel (*Sciurus griseus*) is found in central and western regions of both states; cottontail fore prints rarely register side by side; chipmunk and flying squirrel tracks show a similar pattern, but have a smaller straddle and smaller prints.

 103

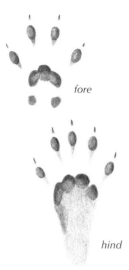

fore

hind

Fore Print
Length: 0.8–1 in (2–2.5 cm)
Width: 0.4–0.8 in (1–2 cm)

Hind Print
Length: 0.7–1.3 in (1.8–3.3 cm)
Width: 0.5–0.9 in (1.3–2.3 cm)

Straddle
2–3.1 in (5.1–7.9 cm)

Stride
Running: 7–15 in (18–38 cm)

Size
Length with tail: 7–9 in (18–23 cm)

Weight
1–2.5 oz (28–71 g)

running

YELLOW-PINE CHIPMUNK
Tamias amoenus

This colorful chipmunk is found in various habitats, from dry sagebrush to coniferous forests, especially where there is yellow pine, in all but the western parts of both states. You are more likely to see or hear this rodent, which is highly active during summer months, than to find its tracks. It is quite happy to climb trees and extract its favorite seeds from ripe pine cones. Chipmunks hibernate in winter, waking up from time to time to have a meal.

Chipmunks are so light that their tracks rarely show fine details. The fore feet each have four toes, while the hind feet have five. These chipmunks run on their toes, so the two heel pads of the fore feet seldom register, and they are completely lacking on the hind feet. The erratic tracks show the hind prints registering ahead of the fore prints. A chipmunk's trail often leads to extensive burrows.

Similar Species: Least Chipmunks (*Tamias minimus*) are smaller and less widespread, being found in southeastern Washington and into Oregon; Townsend's Chipmunks (*T. townsendii*) are found in the western parts of both states; Red-tailed Chipmunks (*T. ruficaudus*) are found in the extreme northeast of Washington; Red Squirrel tracks are larger; mice tracks are smaller; tracks in mid-winter are more likely to belong a squirrel.

fore

hind

Fore Print
Length: 0.5 in (1.3 cm)
Width: 0.5 in (1.3 cm)

Hind Print
Length: 0.6 in (1.5 cm)
Width: 0.5–0.8 in (1.3–2 cm)

Straddle
1.3–2 in (3.3–5.1 cm)

Stride
Walking: 0.8 in (2 cm)
Running/Hopping:
    2–6 in (5.1–15cm)

Size
Length with tail:
    5.5–8 in (14–20 cm)

Weight
0.5–6 oz (14–170 g)

walking

*running
(in snow)*

MEADOW VOLE
(Field Mouse)
Microtus pennsylvanicus

A positive identification for a vole track is next to impossible, because, as with mice, there are so many species to choose from. The Meadow Vole is commonly found in many damp or wet habitats in northern and central of Washington—one possible way to differentiate it from other vole species.

Vole tracks show four toes on the fore print and five on the hind print, although it is hard to tell because the prints are seldom clear. The vole's walk leaves a paired alternating track, with the hind print occasionally registering on the fore. Voles usually opt for faster leaps, however, with the prints appearing in hopping pairs, hind registering on fore. Voles stay under the snow in winter, so look for distinctive piles of cut grass from their ground nests. The bark at the base of shrubs may show tiny teeth marks from gnawing. In summer, voles use the same paths so often that they appear as little runways through the grass.

Similar Species: Montane Voles (*Microtus montanus*) are commonly found at higher elevations in eastern Washington and most of Oregon; Townsend's Voles (*M. townsendii*) and Creeping Voles (*M. oregoni*) are found in the western parts of both states; Deer Mouse tracks show a paired hop pattern with shorter strides.

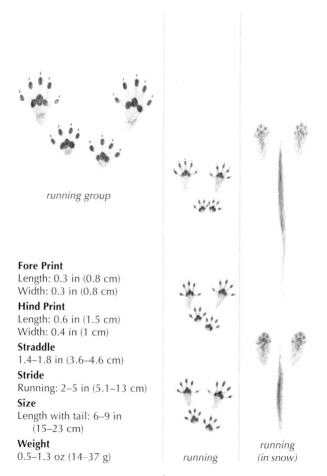

running group

Fore Print
Length: 0.3 in (0.8 cm)
Width: 0.3 in (0.8 cm)

Hind Print
Length: 0.6 in (1.5 cm)
Width: 0.4 in (1 cm)

Straddle
1.4–1.8 in (3.6–4.6 cm)

Stride
Running: 2–5 in (5.1–13 cm)

Size
Length with tail: 6–9 in
(15–23 cm)

Weight
0.5–1.3 oz (14–37 g)

running

*running
(in snow)*

DEER MOUSE
Peromyscus maniculatus

The Deer Mouse is
the most abundant mammal
in the mountains, but it is seldom
seen because it is nocturnal. This
highly adaptable rodent lives from arid
valleys all the way up to alpine meadows,
and may enter buildings to stay active during
winter months.

The fore prints each show four toes, three palm pads
and two heel pads. The hind prints show five toes and
three palm pads, and the heel pads rarely register. It takes
perfect, soft mud to get clear prints from such a tiny
mammal. Running tracks are most noticeable in snow, and
they show the hind prints falling in front of the fore prints,
which are closely set to each other. In soft snow, the prints
may merge to appear as larger pairs of prints with tail drag
evident. Follow the tracks, and they may take you up the
occasional tree or down into a burrow.

Similar Species: Many less common species of mice are
identical; House Mice (*Mus musculus*) are very similar but
associate more with humans; voles have a longer stride;
chipmunks have a wider straddle; jumping mouse prints
may be similar in size; shrews have a narrower straddle.

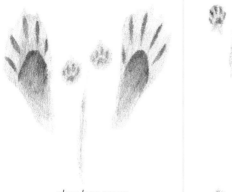

slow hop group

Hind Print
(fore print is much smaller)
Length: 1.5–1.8 in (3.8–4.6 cm)
Width: 0.5–0.8 in (1.3–2 cm)

Straddle
1.3–2.3 in (3.3–5.8 cm)

Stride
Hopping: 5–24 in (13–61 cm)

Size
Length with tail: 8–13.5 in (20–34 cm)

Weight
1.5–2.5 oz (43–71 g)

fast hop

ORD'S KANGAROO RAT
Dipodomys ordii

This small, athletic rodent is capable of big jumps, as its name suggests. The best place to look for the nocturnal Ord's Kangaroo Rat is in the arid, shrubby regions of eastern Oregon, in sandy, dry soils. In cold weather its stays under the snow, venturing out on milder nights.

Because of the kangaroo rat's preference for drier terrain, good tracks are hard to find. Sometimes an abundance of them can be found in sand, but the finer print detail will be lacking. The best way to identify the track is by its habit. When a kangaroo rat hops slowly, the two small fore feet register between the large hind feet, which show a long heel mark. The long tail leaves a dragline. At speed, however, the fore feet do not register, the hind heel appears shorter and the tail infrequently registers. If you find a good trail, it might lead you to this rodent's large nesting mounds. If you find a burrow, tap your fingers by it: you may be surprised to hear something thumping back.

Similar Species: Western Jumping Mice prefer lush areas.

running group

Fore Print
Length: 0.3–0.5 in (0.8–1.3 cm)
Width: 0.3–0.5 in (0.8–1.3 cm)

Hind Print
Length: 0.5–1.3 in (1.3–3.3 cm)
Width: 0.5 in (1.3 cm)

Straddle
1.8–1.9 in (4.6–4.8 cm)

Stride
Hopping: 2–7 in (5.1–18 cm)
In alarm: 3–4 ft (91–122 cm)

Size
Length with tail: 7–9 in (18–23 cm)

Weight
0.6–1.3 oz (17–37 g)

running

WESTERN JUMPING MOUSE
Zapus princeps

Congratulations if you find and successfully identify the tracks of a jumping mouse! These rodents are hard to find, although they are distributed widely in eastern Oregon, often in tall grass meadows. The preference of Western Jumping Mice for grassy meadows, and their long, deep hibernation in winter (about six months!) make tracking very difficult.

Jumping mouse tracks are distinctive if you do find them. The two smaller fore prints register between the long hind prints. The long heels do not always show. When jumping, these mice make short leaps, and the tail may show a dragline in soft mud or unseasonable snow. Clusters of cut grass stems, about 5 inches (13 cm) long, lying in meadows are a more abundant sign of this rodent.

Similar Species: Pacific Jumping Mice (*Zapus trinotatus*) are found in the western parts of both states; Deer Mice have a similar-sized straddle; kangaroo rats also jump, but usually on just their two hind feet, not all four.

running group

Fore Print
Length: 0.2 in (0.5 cm)
Width: 0.2 in (0.5 cm)

Hind Print
Length: 0.6 in (1.5 cm)
Width: 0.3 in (0.8 cm)

Straddle
0.8–1.3 in (2–3.3 cm)

Stride
Running: 1.2–2 in (3–5.1 cm)

Size
Length with tail: 3–5 in (7.6–13 cm)

Weight
0.1–1 oz (3–28 g)

running

DUSKY SHREW
Sorex monticolus

While many species of tiny, frenetic shrews are found in Washington and Oregon, the most likely candidate is the widespread and adaptable Dusky Shrew. This shrew is just as happy in the high heathlands as in the lowland swamps. Its rapid activity makes it difficult to observe closely.

In its energetic and unending quest for food, a shrew usually leaves a running pattern of four prints, but it may slow to an alternating walking pattern. In deeper snow its tail often leaves a dragline. The individual prints in a group are often indistinct, but in shallow, wet snow or mud, even the toes can be counted, with five on both fore and hind feet. If you follow the shrew's trail it may disappear down a burrow. In snow, its tunneling activity may leave a ridge of snow on the surface.

Similar Species: Wandering Shrews (*Sorex vagrans*) prefers moist habitats; Water Shrews (*S. palustris*) are larger and are often found near cold mountain streams; Masked Shrews (*S. cinereus*) are nocturnal and more secretive, and they are found in Washington only; mice show just four toes on each of their fore prints.

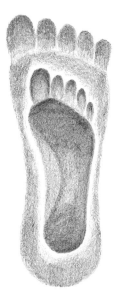

Foot Print
Length: average 14–17 in (36–43 cm)
Width: up to 7 in (18 cm)
Hand Print
Length: up to 12 in (30 cm)
Width: up to 7 in (18 cm)
Stride
Walking: 2–3 ft (61–91 cm)
Size (male>female)
Height: 6–8 ft (1.8–2.4 m)
Weight
400–1000 lb (180–450 kg)

SASQUATCH
Homo cascadensis

Imagine the thrill of being one of the few people to find footprints of the reclusive Sasquatch! Bear prints can be impressively huge, and they are sometimes mistaken for Sasquatch prints, so if you have found large tracks of undetermined origin, study them closely. You will notice many distinctions between a bear's track and a track potentially made by a Sasquatch, such as prominent claws on the bear prints.

This elusive, human-like inhabitant of the remote wilderness has an enormous print that is much broader and more flat-footed than our own—notice the raised arch on the human print. Most Sasquatch tracks have been found along rivers, but because a Sasquatch can weigh so much, a good, clear print can be seen on firmer surfaces. Tracks fall in a pattern just like our own when walking. Juvenile Sasquatch prints might be hard to distinguish from large human footprints. If you think you have found a naked print of huge dimensions, be sure to let somebody know!

Note that a naked human footprint, with which most of us should be familiar, is best seen on sandy beaches near water. There, humans can be seen lying around in large numbers, like lazy sea lions on the coast. Beyond the sandy beaches, however, the human print is usually shod to protect a person's sensitive soles from sticks and stones. Irresponsible humans leave plenty of evidence of their presence on trails. If a woodrat hasn't beaten you to it, do the beautiful mountain wilderness a big favor and pick up the garbage.

BIRDS AND AMPHIBIANS

A guide to animal tracks is not complete without some consideration of birds and amphibians.

Birds come in all shapes and sizes; often there are very subtle differences between species that are not reflected in their tracks. The few birds chosen will demonstrate some of the major differences of birds common to this region.

Of course, many birds spend very little time on the ground, perhaps preferring to be in the trees or just soaring through the skies. Some species, on the other hand, spend a lot of time on the ground, and following their tracks can be entertaining. They can spin around in circles and lead you in all directions. Perhaps the track will suddenly end as the bird takes flight, or maybe it will end as a pile of feathers—the bird having met its grizzly end at the claws, teeth or talons of a hungry predator. A most reliable source for bird tracks is the shores of streams and lakes, where mud can hold a print clearly for a long time. Many species of shorebirds and waterfowl pick their way along through this mud, and the sheer number of tracks can be astonishing.

Many amphibians depend on moist environments; look in the soft mud along the shores of lakes and ponds for their distinctive tracks. Generally, frogs have a different approach to getting around from toads, so it is at least possible to tell these two types of tracks apart; it is very difficult to differentiate between the individual species. In drier environments, most amphibians are ousted by reptiles, which thrive in dryness. Unfortunately, reptiles seldom leave a good track for us to identify, because of this preference for dry terrain.

Print
Length: up to 1.5 in (3.8 cm)
Straddle
1–1.5 in (2.5–3.8 cm)
Stride
Hopping: 1.5–5 in (3.8–13 cm)
Size
5.5–6.5 in (14–17 cm)

DARK-EYED JUNCO
Junco hyemalis

This small and common bird typifies the many small hopping birds found in this region. It has three forward-pointing toes and one longer toe at the rear. The best prints are left in snow, although in deep snow the toe detail is lost, and the feet may show some dragging between the hops.

A good place to study these types of prints is near a birdfeeder. Watch the birds scurry around as they pick up fallen seeds, then have a look at the prints they have left behind. For example, juncos are attracted to seeds that are scattered by chickadees when those birds forage for sunflower seeds in the birdfeeder.

Similar Species: The exact dimensions of the toes may indicate what kind of bird you are tracking–larger birds will have larger footprints; not all birds are present year-round, so keep in mind what the season is, too.

Print
Length: up to 4 in (10 cm)
Straddle
Up to 4 in (10 cm)
Stride
Walking: up to 6 in (15 cm)
Size
24 in (61 cm)

COMMON RAVEN
Corvus corax

This legendary bird spends a lot of time strutting around on the ground—confident behavior that may hint at its intelligence.

Ravens show a typical alternating track, with three thick toes pointing forward and one toe pointing back. When a Raven is in need of greater speed, perhaps for takeoff, it leaves a trail of diagonally placed pairs of prints that are rather irregular.

Similar Species: Other corvids leave similar, but smaller, tracks and also spend a lot of time poking around on the ground; their strides are correspondingly shorter; Crow (*Corvus brachyrhyncos*) prints are up to 3 inches (7.6 cm) long; Black-billed Magpie (*Pica pica*) prints are up to 2 inches (5.1 cm) long.

Print
Length: 2–3 in (5.1–7.6 cm)
Straddle
2–3 in (5.1–7.6 cm)
Stride
Walking: 3–6 in (7.6–15 cm)
Size
15–19 in (38–48 cm)

RUFFED GROUSE
Bonasa umbellus

This ground-dweller prefers the quiet seclusion of coniferous forests in winter, so that will be the best place to find its tracks. Follow them carefully, and you might be startled when the grouse bursts from its cover underneath your feet. Its excellent camouflage usually affords it good protection.

The three, thick front toes leave very clear impressions, but the short rear toe, which is angled off to one side, will not always show up so well. The prints make up a neat, straight track that may reflect this bird's cautious approach to life on the forest floor.

Similar Species: White-tailed Ptarmigans (*Lagopus lecurus*), which are found only in the Cascades, and Blue Grouse (*Dendragapus obscurus*) leave similar tracks, but their prints may be obscured and enlarged by the winter feathers they grow on their feet.

GREAT HORNED OWL
Bubo virginianus

This wide-ranging owl is often seen resting quietly in trees during the day; it prefers to hunt at night. The mark left by this predator can be quite a sight if it registers well. You might stumble across this owl 'strike' and guess that the owl's target could have been a vole scurrying around underneath the snow. If you are a really lucky tracker, you will be following the surface track of an animal that abruptly ends with this strike mark–the animal has been attacked by an owl.

The owl strikes through the snow with its talons, leaving an untidy hole, which is occasionally surrounded by imprints of wing and tail feathers. These feather imprints are made as the owl struggles to take off with possibly heavy prey. The owl is an accomplished hunter in snow, although it is not the most graceful of walkers, and it prefers to fly away from the scene.

Similar Species: Ravens can also leave a strike mark, but usually with much sharper feather imprints.

Print
Length: 4–5 in (10–13 cm)

Straddle
5–7 in (13–18 cm)

Stride
Walking: 5–7 in (13–18 cm)

Size
32–48 in (81–122 cm)

CANADA GOOSE
Branta canadensis

    This common goose is a familiar sight in open areas by lakes and ponds. Its huge webbed feet leave prints that can often be seen in abundance along the muddy shores of just about any waterbody, including in urban parks. Goose droppings can accumulate in prolific amounts.

    The webbed foot has three long toes, all facing forward. These toes register well, but the webbing between them does not always show on the print. The prints point inwards, giving the bird a pigeon-toed appearance, which perhaps accounts for the bird's waddling gait.

Similar Species: These webbed prints are typical of many waterfowl, most of which are smaller, such as the many species of gulls and ducks. If you come across exceptionally large prints similar in shape, they are likely from a swan.

Print
Length: 0.8–1.3 in (2–3.3 cm)
Stride
Erratic
Size
7–8 in (18–20 cm)

SPOTTED SANDPIPER

Actitis macularia

The bobbing tail of this sandpiper is a common sight on the shores of lakes, rivers and streams, but you will usually only find one bird in any given location.

As it teeters up and down on shores, it leaves trails of three-toed prints. Its fourth toe is very small and faces off to one side at an angle. Sandpiper tracks can have an erratic stride.

Similar Species: These tracks are quite typical of all sandpipers and plovers, although there will be much diversity in size.

Print
Length: 1.5 in (3.8 cm)
Stride
1.3 in (3.3 cm)
Size
10.5–11.5 in (27–29 cm)

COMMON SNIPE
Gallinago gallinago

This short-legged character is a resident of marshes and bogs, where its neat prints can often be seen in mud. Snipes are quite secretive when on the ground, and you may be surprised as it suddenly flushes out from beneath your feet. Watch out for the occasional snipe perched on a snag or fencepost, or listen for the eerie whistle as it dives from the sky.

The neat prints show four toes, with the small rear toe pointing inwards. The Common Snipe has a very short stride, because of its short legs and stocky body.

Print
Length: up to 6.5 in (17 cm)
Stride
9 in (23 cm)
Size
50–54 in (127–137 cm)

GREAT BLUE HERON
Ardea herodias

The refined and graceful image of this large heron symbolizes the precious wetlands in which it patiently hunts for food. Still and statuesque as it waits for a meal to swim by, this heron will have cause to walk from time to time, perhaps to find a better hunting location. Look for the large, slender tracks along the banks or mudflats of waterbodies.

A bird that lives and hunts with such precision walks in a similar fashion, leaving straight tracks that nearly fall in a straight line. Look for the slender rear toe in the print.

Similar Species: Cranes often inhabit similar habitats and have similarly sized prints, but their smaller hind toes do not register.

FROGS AND TOADS

The best place to look for frog and toad tracks is undoubtedly along the muddy fringes of waterbodies, but toad tracks can occasionally be found in drier areas, for example as unclear trails in dusty patches of soil. Generally, frogs hop and toads walk, but toads are pretty capable hoppers, too, especially when being hassled by over-enthusiastic naturalists.

FROGS

Among the frog tracks you might find are those of the small Pacific Treefrog (*Hyla regilla*), which is common throughout both states. The much larger Red-legged Frog (*Rana aurora*) is confined to the western fringes of both states, while the hardy Spotted Frog (*R. pretiosa*) is more widespread in the central and eastern regions, frequenting cold mountain streams and lakes. If the track is unusually large, then it is surely from a Bullfrog (*R. catesbeiana*). Up to 8 inches (20 cm) in length, this introduced giant is reportedly spreading its range through most parts of both states.

Hopping action is well displayed by all of these frogs, with the two small fore prints registering in front of the long-toed hind prints. Such tracks may be up to 3 inches (7.6 cm) in straddle, depending on age and species.

TOADS

The toad most likely to be found is the Western Toad (*Bufo boreas*), also called the Boreal Toad, which inhabits streams, meadows and woodlands. It is widely distributed throughout Washington and Oregon. The Great Basin Spadefoot (*Spea intermontana*) is found in the eastern parts of both states. It has a tendency to burrow into sandy soils, out of sight.

Toads leave rather abstract prints as they walk. The heels of the hind feet do not register. In less firm surfaces, you can often see the draglines left by the toes. The straddle may be up to 2.5 inches (6.4 cm).

TRACK PATTERNS & PRINTS

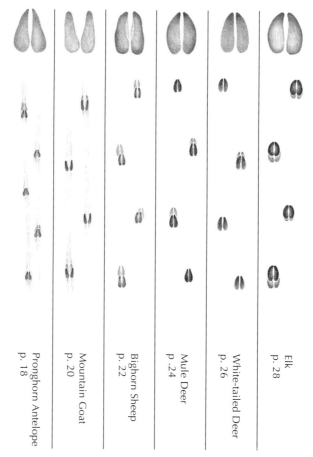

Pronghorn Antelope
p. 18

Mountain Goat
p. 20

Bighorn Sheep
p. 22

Mule Deer
p. 24

White-tailed Deer
p. 26

Elk
p. 28

Horse
p. 30

Black Bear
p. 32

Grizzly Bear
p. 34

Gray Wolf
p. 36

Coyote
p. 38

Gray Fox
p. 40

141

TRACK PATTERNS & PRINTS

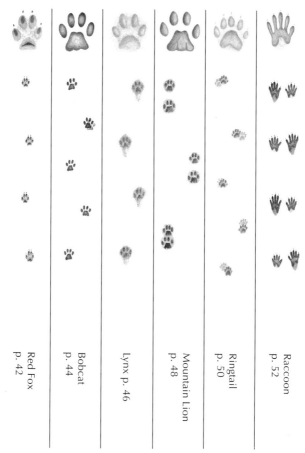

Red Fox
p. 42

Bobcat
p. 44

Lynx p. 46

Mountain Lion
p. 48

Ringtail
p. 50

Raccoon
p. 52

River Otter
p. 54

Fisher
p. 56

Marten
p. 58

Short-tailed Weasel
p. 60

Long-tailed Weasel
p. 62

Wolverine
p. 64

143

TRACK PATTERNS & PRINTS

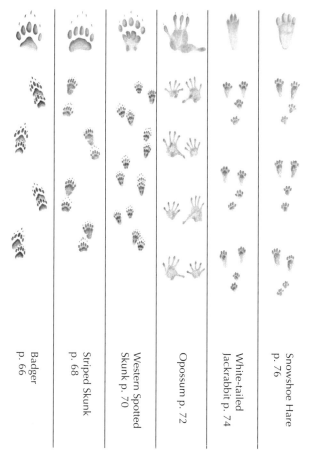

Badger
p. 66

Striped Skunk
p. 68

Western Spotted
Skunk p. 70

Opossum p. 72

White-tailed
Jackrabbit p. 74

Snowshoe Hare
p. 76

Mountain Beaver
p. 88

Muskrat
p. 86

Beaver
p. 84

Porcupine
p. 82

Pika
p. 80

Mountain Cottontail
p. 78

TRACK PATTERNS & PRINTS

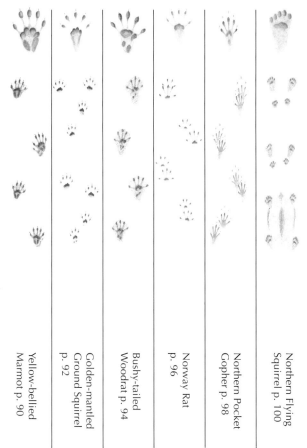

Yellow-bellied
Marmot p. 90

Golden-mantled
Ground Squirrel
p. 92

Bushy-tailed
Woodrat p. 94

Norway Rat
p. 96

Northern Pocket
Gopher p. 98

Northern Flying
Squirrel p. 100

Western Jumping
Mouse p. 112

Ord's Kangaroo Rat
p. 110

Deer Mouse
p. 108

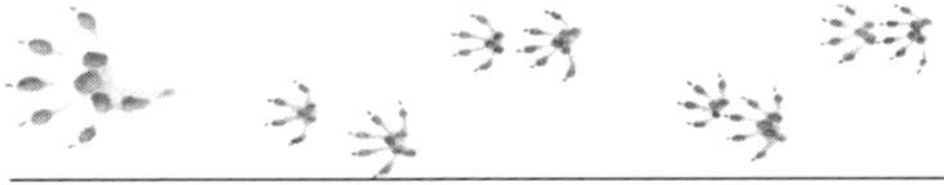

Meadow Vole
p. 106

Yellow-pine
Chipmunk p. 104

Red Squirrel
p. 102

TRACK PATTERNS & PRINTS

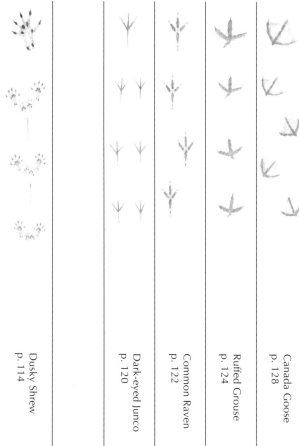

Dusky Shrew
p. 114

Dark-eyed Junco
p. 120

Common Raven
p. 122

Ruffed Grouse
p. 124

Canada Goose
p. 128

Toads p. 138

Frogs p. 136

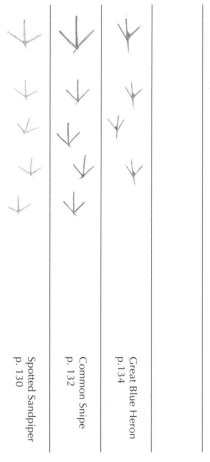

FORE PRINTS

Gray Fox

Red Fox

Bobcat

Coyote

Gray Wolf

Canada Lynx

Mountain Lion

Black Bear

Grizzly Bear

inch   cm
0   0

1

2   5

150

FORE PRINTS

Long-tailed Weasel

Short-tailed Weasel

Ringtail

Striped Skunk

Western
Spotted Skunk

Fisher

River Otter

Marten

Wolverine

Badger

inch   cm
0   0

1

2   5

 151

HOOFED PRINTS

White-tailed Deer

Mule Deer

Pronghorn
Antelope

Bighorn Sheep

Mountain Goat

Elk

inch   cm
0 ————— 0

1

2 ————— 5

Horse

HIND PRINTS

Opossum

Muskrat

Yellow-bellied
Marmot

White-tailed
Jackrabbit

Snowshoe Hare

Mountain
Cottontail

Raccoon

Mountain
Beaver

Beaver

Porcupine

153

HIND PRINTS

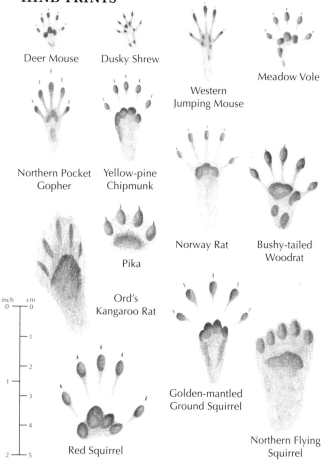

Deer Mouse

Dusky Shrew

Western
Jumping Mouse

Meadow Vole

Northern Pocket
Gopher

Yellow-pine
Chipmunk

Norway Rat

Bushy-tailed
Woodrat

Pika

Ord's
Kangaroo Rat

inch   cm
0   0

1

1   2

3

4

2   5

Red Squirrel

Golden-mantled
Ground Squirrel

Northern Flying
Squirrel

BIBLIOGRAPHY

Barwise, J. E. 1989. *Animal Tracks of Western Canada*. Edmonton, Alberta: Lone Pine Publishing.

Behler, J. L., and F. W. King. 1979. *Field Guide to North American Reptiles and Amphibians*. National Audubon Society. New York: Alfred A. Knopf.

Burt, W. H. 1976. *A Field Guide to the Mammals*. Boston: Houghton Mifflin Company.

Corkran, C. C., and C. Thoms. 1996. *Amphibians of Oregon, Washington and British Columbia*. Edmonton, Alberta: Lone Pine Publishing.

Farrand, J., Jr. 1995. *Familiar Animal Tracks of North America*. National Audubon Society Pocket Guide. New York: Alfred A. Knopf.

Forrest, L. R. 1988. *Field Guide to Tracking Animals in Snow*. Harrisburg: Stackpole Books.

Gadd, B. 1995. *Handbook of the Canadian Rockies*. Jasper, Alberta: Corax Press.

Halfpenny, J. 1986. *A Field Guide to Mammal Tracking in North America*. Boulder: Johnson Publishing Company.

Headstrom, R. 1971. *Identifying Animal Tracks*. Toronto: General Publishing Company.

Kavanagh, J. 1993. *Nature BC*. Edmonton, Alberta: Lone Pine Publishing.

Murie, O. J. 1974. *A Field Guide to Animal Tracks*. The Peterson Field Guide Series. Boston: Houghton Mifflin Company.

Rezendes, P. 1992. *Tracking and the Art of Seeing: How to Read Animal Tracks and Sign*. Vermont: Camden House Publishing.

Scotter, G. W., and T. J. Ulrich. 1995. *Mammals of the Canadian Rockies*. Saskatoon: Fifth House.

Stall, C. 1989. *Animal Tracks of the Rocky Mountains*. Seattle: The Mountaineers.

Stokes, D. and L. Stokes. 1986. *A Guide to Animal Tracking and Behaviour*. Toronto: Little, Brown and Company.

Wassink, J. L. 1993. *Mammals of the Central Rockies*. Missoula: Mountain Press Publishing Company.

Whitaker, J. O., Jr. 1996. *National Audubon Society Field Guide to North American Mammals*. New York: Alfred A. Knopf.

INDEX

Page numbers in **boldface** type refer to the primary (illustrated) treatments of animal species and their tracks.